The Art of Reasoning

Readings for Logical Analysis

The Art of Reasoning

Readings for Logical Analysis

STEPHEN R. C. HICKS
AND DAVID KELLEY

W · W · NORTON & COMPANY
NEW YORK · LONDON

Copyright © 1994 W. W. Norton & Company

Pages 317–20 constitute an extension of the copyright page.

All rights reserved
Printed in the United States of America
Second Expanded Edition

Text of this book composed in Monotype Walbaum with the display set in Corvenus.
Composition by ComCom.
Manufacturing by Haddon Craftsmen.
Book design by Jack Meserole/Antonina Krass.

Library of Congress Cataloging-in-Publication Data

The Art of reasoning : readings for logical analysis / [edited] by
    David Kelley and Stephen R.C. Hicks.
        p.   cm.
    Readings to accompany: The art of reasoning / David Kelley. 2nd
ed. 1994.
    1. Reasoning.   2. Logic.   I. Kelley, David.   II. Hicks, Stephen
Ronald Craig, 1960–   .   III. Kelley, David.   Art of reasoning.
BC177.A78   1994
160—dc20   93-45508

ISBN 0-393-96500-7
W. W. Norton & Company, Inc.
500 Fifth Avenue, New York, N. Y. 10110

W. W. Norton & Company Ltd.
10 Coptic Street, London WC1A 1PU

1 2 3 4 5 6 7 8 9 0

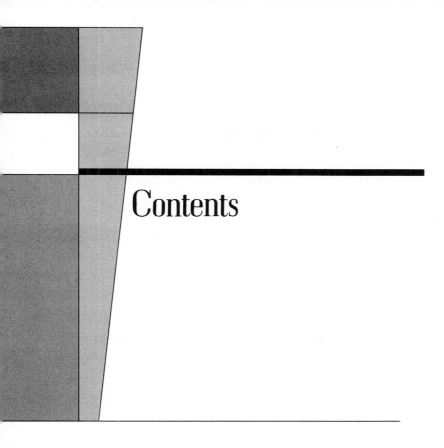

Contents

PART TWO   ON ADULTERY AND MURDER

PART THREE   ON POETRY, PROVERBS, AND CHILDISH INSULTS

PART FOUR   ON EUTHANASIA AND ABORTION

PART FIVE   ON DINOSAURS, PERCEPTION, AND TABLE MANNERS

PART SIX   ON EDUCATION

PART SEVEN   ON JUSTICE AND RIGHTS

PART EIGHT   ON FREEDOM OF SPEECH, ECONOMIC FREEDOM, AND GOVERNMENT REGULATION

PART NINE   ON THE EXISTENCE OF GOD

PART TEN   POVERTY, CRIME, AND GUN CONTROL

Contents by Logical Type

CONCEPTS AND PROPOSITIONS

INDUCTIVE LOGIC

PREFACE

An important need in critical thinking or logic courses is for a variety of examples of real-life argument. Textbooks designed for such courses typically include short examples to illustrate specific points. But many instructors recognize the need for longer passages that call on a wider variety of logical skills. *The Art of Reasoning: Readings for Logical Analysis* is designed to fill that need.

The essays in this reader were selected by the following criteria:

1. All of the major types of argument should be represented. For example, arguments relying heavily on analogies, statistical evidence, deduction, Mill's Methods, definitions, and so on, should be included.
2. The selections should vary in length and degree of difficulty. All, however, should be longer than the examples usually found in logic textbooks. Lengths should range from a couple of hundred words (roughly "letters to the editor" length) to two or three thousand words. In some cases the structure of the argument should be transparent. Others should involve detecting assumed premises, separating compound arguments, eliminating repetition, extracting the argument from background material, and connecting points made in different parts of the essay before the overall theme emerges.

3. The selections should be drawn from a wide range of topics. Examples should include the law, the natural sciences, the social sciences, ethics, history, politics, medicine, religion, and philosophy. This will give students a sense for how the same type of argument can be used in different areas of investigation, as well as how such areas sometimes use specialized argumentative techniques.

4. All of the selections should have been published before, and they should either be by well-known, important, or controversial writers, or be about important or controversial topics. Students then will be dealing with real arguments by real people on real issues, not just examples made up as "academic exercises."

5. Examples of both good and bad argumentation should be included. This is in contrast to the "spot the error" approach that characterizes many textbooks, since the examples included there are overwhelmingly selected as examples of mistakes.

6. More selections should be included than can be used in one course, thus giving the instructor the flexibility to vary the selection used from semester to semester.

7. Finally, to the extent that the selections deal with controversial and sensitive issues, the text should be balanced, to reflect major positions across the spectrum.

Our reader is designed with these considerations in mind.

We have included two tables of contents. The first classifies the selections on the basis of subject matter; the second on the basis of the logical techniques they illustrate. In addition, each selection is accompanied by study questions asking the student to identify key logical elements in the passage.

We would like to thank Rockford College for a Mary Ashby Cheek grant to help defray research expenses; James Kitner and Dena Mattausch-Hicks for research and editorial assistance; and Roby Harrington, Allen Clawson, and Nancy Yanchus of W. W. Norton for their professional and friendly guidance throughout the project.

The Art of Reasoning

Readings for Logical Analysis

PART ONE

On Torture and the Death Penalty

MICHAEL LEVIN

The Case for Torture

Michael Levin, author of Feminism and Freedom *and* Meta-physics and the Mind-Body Problem, *is a professor of philosophy at the City University of New York.*

It is generally assumed that torture is impermissible, a throwback   1
to a more brutal age. Enlightened societies reject it outright, and
regimes suspected of using it risk the wrath of the United States.

I believe this attitude is unwise. There are situations in which   2
torture is not merely permissible but morally mandatory. Moreover,
these situations are moving from the realm of imagination to fact.

Death: Suppose a terrorist has hidden an atomic bomb on Man-   3

[Michael Levin, "The Case for Torture." *Newsweek*, June 7, 1982, p. 13.]

hattan Island which will detonate at noon on July 4 unless . . . (here follow the usual demands for money and release of his friends from jail). Suppose, further, that he is caught at 10 a.m. of the fateful day, but—preferring death to failure—won't disclose where the bomb is. What do we do? If we follow due process—wait for his lawyer, arraign him—millions of people will die. If the only way to save those lives is to subject the terrorist to the most excruciating possible pain, what grounds can there be for not doing so? I suggest there are none. In any case, I ask you to face the question with an open mind.

4   Torturing the terrorist is unconstitutional? Probably. But millions of lives surely outweigh constitutionality. Torture is barbaric? Mass murder is far more barbaric. Indeed, letting millions of innocents die in deference to one who flaunts his guilt is moral cowardice, an unwillingness to dirty one's hands. If *you* caught the terrorist, could you sleep nights knowing that millions died because you couldn't bring yourself to apply the electrodes?

5   Once you concede that torture is justified in extreme cases, you have admitted that the decision to use torture is a matter of balancing innocent lives against the means needed to save them. You must now face more realistic cases involving more modest numbers. Someone plants a bomb on a jumbo jet. He alone can disarm it, and his demands cannot be met (or if they can, we refuse to set a precedent by yielding to his threats). Surely we can, we must, do anything to the extortionist to save the passengers. How can we tell 300, or 100, or 10 people who never asked to be put in danger, "I'm sorry, you'll have to die in agony, we just couldn't bring ourselves to . . ."

6   Here are the results of an informal poll about a third, hypothetical, case. Suppose a terrorist group kidnapped a newborn baby from a hospital. I asked four mothers if they would approve of torturing kidnappers if that were necessary to get their own newborns back. All said yes, the most "liberal" adding that she would like to administer it herself.

7   I am not advocating torture as punishment. Punishment is addressed to deeds irrevocably past. Rather, I am advocating torture as an acceptable measure for preventing future evils. So understood, it is far less objectionable than many extant punishments. Opponents of the death penalty, for example, are forever insisting that executing a murderer will not bring back his victim (as if the

purpose of capital punishment were supposed to be resurrection, not deterrence or retribution). But torture, in the cases described, is intended not to bring anyone back but to keep innocents from being dispatched. The most powerful argument against using torture as a punishment or to secure confessions is that such practices disregard the rights of the individual. Well, if the individual is all that important—and he is—it is correspondingly important to protect the rights of individuals threatened by terrorists. If life is so valuable that it must never be taken, the lives of the innocents must be saved even at the price of hurting the one who endangers them.

Better precedents for torture are assassination and pre-emptive attack. No Allied leader would have flinched at assassinating Hitler, had that been possible. (The Allies did assassinate Heydrich.) Americans would be angered to learn that Roosevelt could have had Hitler killed in 1943—thereby shortening the war and saving millions of lives—but refused on moral grounds. Similarly, if nation A learns that nation B is about to launch an unprovoked attack, A has a right to save itself by destroying B's military capability first. In the same way, if the police can by torture save those who would otherwise die at the hands of kidnappers or terrorists, they must. 8

Idealism: There is an important difference between terrorists and their victims that should mute talk of the terrorists' "rights." The terrorist's victims are at risk unintentionally, not having asked to be endangered. But the terrorist knowingly initiated his actions. Unlike his victims, he volunteered for the risks of his deed. By threatening to kill for profit or idealism, he renounces civilized standards, and he can have no complaint if civilization tries to thwart him by whatever means necessary. 9

Just as torture is justified only to save lives (not extort confessions or recantations), it is justifiably administered only to those *known* to hold innocent lives in their hands. Ah, but how can the authorities ever be sure they have the right malefactor? Isn't there a danger of error and abuse? Won't We turn into Them? 10

Questions like these are disingenuous in a world in which terrorists proclaim themselves and perform for television. The name of their game is public recognition. After all, you can't very well intimidate a government into releasing your freedom fighters unless you announce that it is your group that has seized its embassy. "Clear guilt" is difficult to define, but when 40 million people see 11 a group of masked gunmen seize an airplane on the evening news, there is not much question about who the perpetrators are. There will be hard cases where the situation is murkier. Nonetheless, a line demarcating the legitimate use of torture can be drawn. Torture only the obviously guilty, and only for the sake of saving innocents, and the line between Us and Them will remain clear.

12   There is little danger that the Western democracies will lose their way if they choose to inflict pain as one way of preserving order. Paralysis in the face of evil is the greater danger. Some day soon a terrorist will threaten tens of thousands of lives, and torture will be the only way to save them. We had better start thinking about this.

STUDY QUESTIONS

1. Levin argues that torture is justifiable in some cases, but not in others. What conditions does he propose to distinguish the two sorts of cases?

2. Levin's conclusion is that torture is justifiable in some cases. How many different arguments does he use in support of this conclusion?

3. How does Levin respond to the charge that torture would violate the terrorist's rights? Does he think terrorists have rights?

4. What is the point of the "informal poll" Levin reports in paragraph 6? In raising such cases, does Levin run the risk of committing an appeal to emotion?

5. In paragraph 8, Levin compares torture to assassination and preemptive attacks. What points of similarity does he see among the three? Does he overlook any significant differences?

6. Who makes the decision about whether torture is justifiable? In emergency situations, who would decide and ensure that no abuses occur? What if an innocent person is mistakenly tortured? Why, in paragraph 11, does Levin suggest that questions such as these may be "disingenuous"?

Reference Chapters: 6, 7, 11, 16

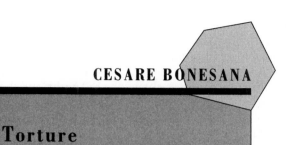

CESARE BONESANA

Torture

*Cesare Bonesana (1738–1794) was the Marchese di Beccarria
and an Italian jurist. The following argument against the use
of torture, especially by legal authorities, is excerpted from his*
Trato dei Delitti e delle Pene *(*On Crimes & Punishments*),
written in 1764.*

A cruelty consecrated among most nations by custom is the torture   1
of the accused during his trial, on the pretext of compelling him to
confess his crime, of clearing up contradictions in his statements, of
discovering his accomplices, of purging him in some metaphysical
and incomprehensible way from infamy, or finally of finding out
other crimes of which he may possibly be guilty, but of which he is
not accused.

A man cannot be called *guilty* before sentence has been passed   2
on him by a judge, nor can society deprive him of its protection till
it has been decided that he has broken the condition on which it was
granted. What, then, is that right but one of mere might by which
a judge is empowered to inflict a punishment on a citizen whilst his
guilt or innocence are still undetermined? The following dilemma
is no new one: either the crime is certain or uncertain; if certain, no
other punishment is suitable for it than that affixed to it by law; and
torture is useless, for the same reason that the criminal's confession

[Cesare Bonesana, "Torture," excerpt from James A. Farrar, ed., *Crimes and
Punishments.* London: Chatto & Windus, 1880, pp. 148–152.]

is useless. If it is uncertain, it is wrong to torture an innocent person, such as the law adjudges him to be, whose crimes are not yet proved.

3   What is the political object of punishments? The intimidation of other men. But what shall we say of the secret and private tortures which the tyranny of custom exercises alike upon the guilty and the innocent? It is important, indeed, that no open crime shall pass unpunished; but the public exposure of a criminal whose crime was hidden in darkness is utterly useless. An evil that has been done and cannot be undone can only be punished by civil society insofar as it may affect others with the hope of impunity. If it be true that there are a greater number of men who either from fear or virtue respect the laws than of those who transgress them, the risk of torturing an innocent man should be estimated according to the probability that any man will have been more likely, other things being equal, to have respected than to have despised the laws.

4   But I say in addition: it is to seek to confound all the relations of things to require a man to be at the same time accuser and accused, to make pain the crucible of truth, as if the test of it lay in the muscles and sinews of an unfortunate wretch. The law which ordains the use of torture is a law which says to men: "Resist pain; and if Nature has created in you an inextinguishable self-love, if she has given you an inalienable right of self-defence, I create in you a totally contrary affection, namely, an heroic self-hatred, and I command you to accuse yourselves, and to speak the truth between the laceration of your muscles and the dislocation of your bones."

5   This infamous crucible of truth is a still-existing monument of that primitive and savage legal system which called trials by fire and boiling water, or the accidental decisions of combat, *judgments of God*, as if the rings of the eternal chain in the control of the First Cause must at every moment be disarranged and put out for the petty institutions of mankind. The only difference between torture and the trial by fire and water is, that the result of the former seems to depend on the will of the accused, and that of the other two on a fact which is purely physical and extrinsic to the sufferer; but the difference is only apparent, not real. The avowal of truth under tortures and agonies is as little free as it was in those times the prevention without fraud of the usual effects of fire and boiling water. Every act of our will is ever proportioned to the force of the sensible impression which causes it, and the sensibility of every man

is limited. Hence the impression produced by pain may be so intense as to occupy a man's entire sensibility and leave him no other liberty than the choice of the shortest way of escape, for the present moment, from his penalty. Under such circumstances the answer of the accused is as inevitable as the impressions produced by fire and water; and the innocent man who is sensitive will declare himself guilty, when by so doing he hopes to bring his agonies to an end. All the difference between guilt and innocence is lost by virtue of the very means which they profess to employ for its discovery.

Torture is a certain method for the acquittal of robust villains and for the condemnation of innocent but feeble men. See the fatal drawbacks of this pretended test of truth—a test, indeed, that is worthy of cannibals; a test which the Romans, barbarous as they too were in many respects, reserved for slaves alone, the victims of their fierce and too highly lauded virtue. Of two men, equally innocent or equally guilty, the robust and courageous will be acquitted, the weak and the timid will be condemned, by virtue of the following exact train of reasoning on the part of the judge: "I as judge had to find you guilty of such and such a crime; you, AB, have by your physical strength been able to resist pain, and therefore I acquit you; you, CD, in your weakness have yielded to it; therefore I condemn you. I feel that a confession extorted amid torments can have no force, but I will torture you afresh unless you corroborate what you have now confessed."

The result, then, of torture is a matter of temperament, of calculation, which varies with each man according to his strength and sensibility; so that by this method a mathematician might solve better than a judge this problem: "Given the muscular force and the nervous sensibility of an innocent man, to find the degree of pain which will cause him to plead guilty to a given crime."

STUDY QUESTIONS

1. In his opening paragraph, Bonesana lists five attempted justifications for using torture. Does he address each of them in the rest of the selection?
2. In paragraph 2, Bonesana poses a dilemma for those advocat-

ing the use of torture. Diagram the structure of the dilemma argument, and then explain why he thinks torture is both unsuitable and useless if the crime is certain.

3. How would Bonesana respond to the following objection: "Suppose the crime is certain, and we know the guilty party had accomplices. You [Bonesana] haven't proved that torturing the guilty party wouldn't give us a chance to catch the accomplices"?

4. What do you think Bonesana would say in response to Levin's argument for torture in some cases?

Reference Chapters: 5, 10

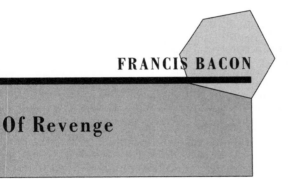

FRANCIS BACON

Of Revenge

Francis Bacon (1560–1626) was educated at Trinity College, Cambridge. He wrote extensively on scientific and philosophical issues and was a member of the English Civil Service, holding at different times the positions of Solicitor-General, Attorney-General, and Lord Chancellor.

1   Revenge is a kind of wild justice, which the more man's nature runs to, the more ought law to weed it out; for as for the first wrong, it doth but offend the law, but the revenge of that wrong putteth the

[Francis Bacon, "Of Revenge," in *Essays: Or Counsels, Civil and Moral*, 12th ed. Boston: Little, Brown, and Co., 1879, pp. 63–64.]

law out of office. Certainly, in taking revenge, a man is but even with his enemy, but in passing it over, he is superior; for it is a prince's part to pardon; and Solomon, I am sure, saith, "It is the glory of a man to pass by an offence." That which is past is gone and irrevocable, and wise men have enough to do with things present and to come; therefore they do but trifle with themselves that labor in past matters. There is no man doth a wrong for the wrong's sake, but thereby to purchase himself profit, or pleasure, or honor, or the like; therefore, why should I be angry with a man for loving himself better than me? And if any man should do wrong merely out of ill-nature, why, yet it is but like the thorn or briar, which prick and scratch, because they can do no other. The most tolerable sort of revenge is for those wrongs which there is no law to remedy; but then, let a man take heed the revenge be such as there is no law to punish, else a man's enemy is still beforehand, and it is two for one. Some, when they take revenge, are desirous the party should know whence it cometh. This is the more generous; for the delight seemeth to be not so much in doing the hurt as in making the party repent; but base and crafty cowards are like the arrow that flieth in the dark. Cosmus, Duke of Florence,[1] had a desperate saying against perfidious or neglecting friends, as if those wrongs were unpardonable. "You shall read," saith he, "that we are commanded to forgive our enemies; but you never read that we are commanded to forgive our friends." But yet the spirit of Job was in a better tune: "Shall we," saith he, "take good at God's hands, and not be content to take evil also?"[2] and so of friends in a proportion. This is certain, that a man that studieth revenge keeps his own wounds green, which otherwise would heal and do well. Public revenges[3] are for the most part fortunate; as that for the death of Caesar;[4] for the death of

1. He alludes to Cosmo de Medici, or Cosmo I., chief of the Republic of Florence, the encourager of literature and the fine arts.

2. Job ii. 10.—"Shall we receive good at the hand of God, and shall we not receive evil?"

3. By "public revenges," he means punishment awarded by the state with the sanction of the laws.

4. He alludes to the retribution dealt by Augustus and Antony to the murderers of Julius Caesar. It is related by ancient historians, as a singular fact, that not one of them died a natural death.

Pertinax; for the death of Henry the Third of France;[5] and many more. But in private revenges it is not so; nay, rather, vindictive persons live the life of witches, who, as they are mischievous, so end they unfortunate.

STUDY QUESTIONS

1. Does Bacon offer a definition of "revenge"?
2. Bacon offers at least six different reasons why one should not pursue revenge. Rank them in order of strength.
3. Bacon claims that "It is the glory of a man to pass by an offence." Does it follow from this principle that we should not have a legal system? Since Bacon clearly thinks there is a role for the law, how do you think he would respond to this?

Reference Chapter: 3

5. Henry III of France was assassinated in 1599, by Jacques Clement, a Jacobin monk, in the frenzy of fanaticism. Although Clement justly suffered punishment, the end of this bloodthirsty and bigoted tyrant may be justly deemed a retribution dealt by the hand of an offended Providence; so truly does the Poet say:—
—"neque enim lex sequior ulla
Quam necis artifices arte perire suâ."
["For there is no law fairer than one by which the killer is destroyed by his own device."—Eds.]

SISSEL SETERAS STOKES

Capital Punishment Is Immoral

The following selection from the letters-to-the-editor column of a newspaper is representative of the style and length of such letters. Since newspaper editors usually set strict word limits, writers must compress their arguments into as little space as possible. Consequently, they sometimes state their arguments as enthymemes, like those supporting Stokes's conclusion below.

To the editor:

I am writing as a member of the international human rights   1
organization Amnesty International to express my anxiety and concern about the practicing of the death penalty in the state of Indiana.

As you are certainly aware, most European countries have abol-   2
ished the death penalty, with Turkey as the only Western European country to have carried out the death sentence in recent years. In Indiana prisons there are even two juvenile offenders on Death Row. Although a bill has been passed raising the minimum age for passing a death sentence from 10 to 16 years of age, it is still in contravention with agreed international standards stating minimum of 18 years of age for offenders at the time of the crime. I therefore welcome a statement issued by the Indiana Council of Churches expressing the following views:

We oppose the death penalty for four reasons: 3

[Sissel Seteras Stokes, from "Letters to the Editor," *Bloomington* (Indiana) *Herald-Telephone*, September 6, 1987.]

1. We consider it to be morally wrong.

2. The death penalty is used discriminately. Almost all the executed are poor. A large percentage belongs to minority groups.

3. There is no evidence proving the death penalty of having a deterrent effect.

4. An execution is irrevocable.

Study Questions

1. Stokes mentions that she is a member of Amnesty International. What role does this play in her argument?

2. Stokes also mentions Western European countries and an international agreement. What role do these play in her argument?

3. Like many letters to the editor, this one contains several enthymemes. Reconstruct Stokes's arguments by filling in the assumed premises.

Reference Chapters: 6, 7

H. L. MENCKEN

The Penalty of Death

Henry Louis Mencken (1880–1956) was an American newspaper reporter, editorialist, and columnist, best known for his salty and boisterous writing style.

Of the arguments against capital punishment that issue from up-lifters, two are commonly heard most often, to wit:

1. That hanging a man (or frying him or gassing him) is a dreadful business, degrading to those who have to do it and revolting to those who have to witness it.
2. That it is useless, for it does not deter others from the same crime.

The first of these arguments, it seems to me, is plainly too weak to need serious refutation. All it says, in brief, is that the work of the hangman is unpleasant. Granted. But suppose it is? It may be quite necessary to society for all that. There are, indeed, many other jobs that are unpleasant, and yet no one thinks of abolishing them—that of the plumber, that of the soldier, that of the garbage-man, that of the priest hearing confessions, that of the sand-hog, and so on. Moreover, what evidence is there that any actual hangman complains of his work? I have heard none. On the contrary, I have known many who delighted in their ancient art, and practiced it proudly.

In the second argument of the abolitionists there is rather more force, but even here, I believe, the ground under them is shaky.

Their fundamental error consists in assuming that the whole aim of punishing criminals is to deter other (potential) criminals—that we hang or electrocute A simply in order to so alarm B that he will not kill C. This, I believe, is an assumption which confuses a part with the whole. Deterrence, obviously, is *one* of the aims of punishment, but it is surely not the only one. On the contrary, there are at least a half dozen, and some are probably quite as important. At least one of them, practically considered, is *more* important. Commonly, it is described as revenge, but revenge is really not the word for it. I borrow a better term from the late Aristotle: *katharsis*. *Katharsis*, so used, means a salubrious discharge of emotions, a healthy letting off of steam. A school-boy, disliking his teacher, deposits a tack upon the pedagogical chair; the teacher jumps and the boy laughs. This is *katharsis*. What I contend is that one of the prime objects of all judicial punishments is to afford the same grateful relief *(a)* to the immediate victims of the criminal punished, and *(b)* to the general body of moral and timorous men.

4   These persons, and particularly the first group, are concerned only indirectly with deterring other criminals. The thing they crave primarily is the satisfaction of seeing the criminal actually before them suffer as he made them suffer. What they want is the peace of mind that goes with the feeling that accounts are squared. Until they get that satisfaction they are in a state of emotional tension, and hence unhappy. The instant they get it they are comfortable. I do not argue that this yearning is noble; I simply argue that it is almost universal among human beings. In the face of injuries that are unimportant and can be borne without damage it may yield to higher impulses; that is to say, it may yield to what is called Christian charity. But when the injury is serious Christianity is adjourned, and even saints reach for their sidearms. It is plainly asking too much of human nature to expect it to conquer so natural an impulse. A keeps a store and has a bookkeeper, B. B steals $700, employs it in playing at dice or bingo, and is cleaned out. What is A to do? Let B go? If he does so he will be unable to sleep at night. The sense of injury, of injustice, of frustration will haunt him like pruritus. So he turns B over to the police, and they hustle B to prison. Thereafter A can sleep. More, he has pleasant dreams. He pictures B chained to the wall of a dungeon a hundred feet underground, devoured by rats and scorpions. It is so agreeable that it makes him forget his $700. He has got his *katharsis*.

The same thing precisely takes place on a larger scale when there   5
is a crime which destroys a whole community's sense of security.
Every law-abiding citizen feels menaced and frustrated until the
criminals have been struck down—until the communal capacity to
get even with them, and more than even, has been dramatically
demonstrated. Here, manifestly, the business of deterring others is
no more than an afterthought. The main thing is to destroy the
concrete scoundrels whose act has alarmed everyone, and thus made
everyone unhappy. Until they are brought to book that unhappiness
continues; when the law has been executed upon them there is a
sigh of relief. In other words, there is *katharsis*.

I know of no public demand for the death penalty for ordinary   6
crimes, even for ordinary homicides. Its infliction would shock all
men of normal decency of feeling. But for crimes involving the
deliberate and inexcusable taking of human life, by men openly
defiant of all civilized order—for such crimes it seems, to nine men
out of ten, a just and proper punishment. Any lesser penalty leaves
them feeling that the criminal has got the better of society—that
he is free to add insult to injury by laughing. That feeling can be
dissipated only by a recourse to *katharsis*, the invention of the
aforesaid Aristotle. It is more effectively and economically achieved,
as human nature now is, by wafting the criminal to realms of bliss.

The real objection to capital punishment doesn't lie against the   7
actual extermination of the condemned, but against our brutal
American habit of putting it off so long. After all, every one of us
must die soon or late, and a murderer, it must be assumed, is one
who makes that sad fact the cornerstone of his metaphysic. But it is
one thing to die, and quite another thing to lie for long months and
even years under the shadow of death. No sane man would choose
such a finish. All of us, despite the Prayer Book, long for a swift and
unexpected end. Unhappily, a murderer, under the irrational
American system, is tortured for what, to him, must seem a whole
series of eternities. For months on end he sits in prison while his
lawyers carry on their idiotic buffoonery with writs, injunctions,
mandamuses, and appeals. In order to get his money (or that of his
friends) they have to feed him with hope. Now and then, by the
imbecility of a judge or some trick of juridic science, they actually
justify it. But let us say that, his money all gone, they finally throw
up their hands. Their client is now ready for the rope or the chair.
But he must still wait for months before it fetches him.

8   That wait, I believe, is horribly cruel. I have seen more than one man sitting in the death-house, and I don't want to see any more. Worse, it is wholly useless. Why should he wait at all? Why not hang him the day after the last court dissipates his last hope? Why torture him as not even cannibals would torture their victims? The common answer is that he must have time to make his peace with God. But how long does that take? It may be accomplished, I believe, in two hours quite as comfortably as in two years. There are, indeed, no temporal limitations upon God. He could forgive a whole herd of murderers in a millionth of a second. More, it has been done.

STUDY QUESTIONS

1. In paragraph 3, Mencken substitutes the word *katharsis* for "revenge." Do the two words stand for the same concept? If so, what does Mencken's argument gain by making the substitution?

2. In paragraph 2, Mencken draws an analogy between the work of the hangman and the work of the plumber and the garbage-man. In what ways are the jobs not analogous? Do these disanalogies weaken Mencken's point?

3. Mencken says in paragraph 3 that there are "at least a half dozen" aims of punishment. How many does he actually discuss?

4. To what cases does Mencken restrict the use of the death penalty? On what grounds does he restrict it to these cases?

5. How do you think Mencken would respond to the following: "In paragraphs 7 and 8, you [Mencken] show great concern for the condemned criminal. Doesn't this contradict your earlier professed willingness to let the criminal suffer so that his or her victims may get their *katharsis?*"

6. Is Mencken arguing that capital punishment is just and ethical, or is he arguing that it is an acceptable response to understandable (though perhaps unworthy) feelings on the part of most people?

Reference Chapters: 3, 7, 16

JUSTICE WILLIAM BRENNAN

Furman v. *Georgia* [Capital Punishment Is Unconstitutional]

*William Brennan, born in 1906, was educated at the Univer-
sity of Pennsylvania and Harvard. He began his career as a
lawyer in New Jersey and in 1956 became a U.S. Supreme
Court Justice. In the following selection from a 1972 Supreme
Court case, Justice Brennan argues that capital punishment
violates the Eighth Amendment to the U.S. Constitution's prohibition of "cruel and unusual" punishment. Brennan has earlier argued that the wording of the Eighth Amendment is not
precise, and so he has offered four principles for determining
whether a punishment is cruel and unusual:*

> *"[A] punishment must not be so severe as to be degrading to
> the dignity of human beings";*
> *"[T]he State must not arbitrarily inflict a severe punishment";*
> *"[A] severe punishment must not be unacceptable to contempo-
> rary society";*
> *"[A] severe punishment must not be excessive."*

*In the selection that follows, Brennan proceeds to argue that
capital punishment violates all four of these principles.*

[Justice Brennan, Section III of concurring opinion in *Furman* v. *Georgia* 408
U.S. 285–305 (1972).]

III

1   . . . The question, then, is whether the deliberate infliction of death is today consistent with the command of the Clause that the State may not inflict punishments that do not comport with human dignity. I will analyze the punishment of death in terms of the principles set out above and the cumulative test to which they lead: It is a denial of human dignity for the State arbitrarily to subject a person to an unusually severe punishment that society has indicated it does not regard as acceptable, and that cannot be shown to serve any penal purpose more effectively than a significantly less drastic punishment. Under these principles and this test, death is today a "cruel and unusual" punishment.

2   Death is a unique punishment in the United States. In a society that so strongly affirms the sanctity of life, not surprisingly the common view is that death is the ultimate sanction. This natural human feeling appears all about us. There has been no national debate about punishment, in general or by imprisonment, comparable to the debate about the punishment of death. No other punishment has been so continuously restricted, nor has any State yet abolished prisons, as some have abolished this punishment. And those States that still inflict death reserve it for the most heinous crimes. Juries, of course, have always treated death cases differently, as have governors exercising their commutation powers. Criminal defendants are of the same view. "As all practicing lawyers know, who have defended persons charged with capital offenses, often the only goal possible is to avoid the death penalty." Some legislatures have required particular procedures, such as two-stage trials and automatic appeals, applicable only in death cases. "It is the universal experience in the administration of criminal justice that those charged with capital offenses are granted special considerations." This Court, too, almost always treats death cases as a class apart. And the unfortunate effect of this punishment upon the functioning of the judicial process is well known; no other punishment has a similar effect.

3   The only explanation for the uniqueness of death is its extreme severity. Death is today an unusually severe punishment, unusual in its pain, in its finality, and in its enormity. No other existing punishment is comparable to death in terms of physical and mental

suffering. Although our information is not conclusive, it appears that there is no method available that guarantees an immediate and painless death. Since the discontinuance of flogging as a constitutionally permissible punishment, death remains as the only punishment that may involve the conscious infliction of physical pain. In addition, we know that mental pain is an inseparable part of our practice of punishing criminals by death, for the prospect of pending execution exacts a frightful toll during the inevitable long wait between the imposition of sentence and the actual infliction of death. As the California Supreme Court pointed out, "the process of carrying out a verdict of death is often so degrading and brutalizing to the human spirit as to constitute psychological torture." Indeed, as Mr. Justice [Felix] Frankfurter noted, "the onset of insanity while awaiting execution of a death sentence is not a rare phenomenon." The "fate of ever-increasing fear and distress" to which the expatriate is subjected can only exist to a greater degree for a person confined in prison awaiting death.

The unusual severity of death is manifested most clearly in its 4
finality and enormity. Death, in these respects, is in a class by itself. Expatriation, for example, is a punishment that "destroys for the individual the political existence that was centuries in the development," that "strips the citizen of his status in the national and international political community," and that puts "[h]is very existence" in jeopardy. Expatriation thus inherently entails "the total destruction of the individual's status in organized society." "In short, the expatriate has lost the right to have rights." Yet, demonstrably, expatriation is not "a fate worse than death." Although death, like expatriation, destroys the individual's "political existence" and his "status in organized society," it does more, for, unlike expatriation, death also destroys "[h]is very existence." There is, too, at least the possibility that the expatriate will in the future regain "the right to have rights." Death forecloses even that possibility.

Death is truly an awesome punishment. The calculated killing 5
of a human being by the State involves, by its very nature, a denial of the executed person's humanity. The contrast with the plight of a person punished by imprisonment is evident. An individual in prison does not lose "the right to have rights." A prisoner retains, for example, the constitutional rights to the free exercise of religion,

to be free of cruel and unusual punishments, and to treatment as a "person" for purposes of due process of law and the equal protection of the laws. A prisoner remains a member of the human family. Moreover, he retains the right of access to the courts. His punishment is not irrevocable. Apart from the common charge, grounded upon the recognition of human fallibility, that the punishment of death must inevitably be inflicted upon innocent men, we know that death has been the lot of men whose convictions were unconstitutionally secured in view of later, retroactively applied, holdings of this Court. The punishment itself may have been unconstitutionally inflicted, yet the finality of death precludes relief. An executed person has indeed "lost the right to have rights." As one 19th century proponent of punishing criminals by death declared, "When a man is hung, there is an end of our relations with him. His execution is a way of saying, 'You are not fit for this world, take your chance elsewhere.'"

6   In comparison to all other punishments today, then, the deliberate extinguishment of human life by the State is uniquely degrading to human dignity. I would not hesitate to hold, on that ground alone, that death is today a "cruel and unusual" punishment, were it not that death is a punishment of long-standing usage and acceptance in this country. I therefore turn to the second principle— that the State may not arbitrarily inflict an unusually severe punishment.

7   The outstanding characteristic of our present practice of punishing criminals by death is the infrequency with which we resort to it. The evidence is conclusive that death is not the ordinary punishment for any crime.

8   There has been a steady decline in the infliction of this punishment in every decade since the 1930's, the earliest period for which accurate statistics are available. In the 1930's, executions averaged 167 per year; in the 1940's, the average was 128; in the 1950's, it was 72; and in the years 1960–1962, it was 48. There have been a total of 46 executions since then, 36 of them in 1963–1964. Yet our population and the number of capital crimes committed have increased greatly over the past four decades. The contemporary rarity of the infliction of this punishment is thus the end result of a long-continued decline. That rarity is plainly revealed by an exami-

nation of the years 1961–1970, the last 10-year period for which statistics are available. During that time, an average of 106 death sentences was imposed each year. Not nearly that number, however, could be carried out, for many were precluded by commutations to life or a term of years, transfers to mental institutions because of insanity, resentences to life or a term of years, grants of new trials and orders for resentencing, dismissals of indictments and reversals of convictions, and deaths by suicide and natural causes. On January 1, 1961, the death row population was 219; on December 31, 1970, it was 608; during that span, there were 135 executions. Consequently, had the 389 additions to death row also been executed, the annual average would have been 52. In short, the country might, at most, have executed one criminal each week. In fact, of course, far fewer were executed. Even before the moratorium on executions began in 1967, executions totaled only 42 in 1961 and 47 in 1962, an average of less than one per week; the number dwindled to 21 in 1963, to 15 in 1964, and to seven in 1965; in 1966, there was one execution, and in 1967, there were two.

When a country of over 200 million people inflicts an unusually   9
severe punishment no more than 50 times a year, the inference is strong that the punishment is not being regularly and fairly applied. To dispel it would indeed require a clear showing of nonarbitrary infliction.

Although there are not exact figures available, we know that   10
thousands of murders and rapes are committed annually in States where death is an authorized punishment for those crimes. However the rate of infliction is characterized—as "freakishly" or "spectacularly" rare, or simply as rare—it would take the purest sophistry to deny that death is inflicted in only a minute fraction of these cases. How much rarer, after all, could the infliction of death be?

When the punishment of death is inflicted in a trivial number of   11
the cases in which it is legally available, the conclusion is virtually inescapable that it is being inflicted arbitrarily. Indeed, it smacks of little more than a lottery system. The States claim, however, that this rarity is evidence not of arbitrariness, but of informed selectivity: Death is inflicted, they say, only in "extreme" cases.

Informed selectivity, of course, is a value not to be denigrated.   12

Yet presumably the States could make precisely the same claim if there were 10 executions per year, or five, or even if there were but one. That there may be as many as 50 per year does not strengthen the claim. When the rate of infliction is at this low level, it is highly implausible that only the worst criminals or the criminals who commit the worst crimes are selected for this punishment. No one has yet suggested a rational basis that could differentiate in those terms the few who die from the many who go to prison. Crimes and criminals simply do not admit of a distinction that can be drawn so finely as to explain, on that ground, the execution of such a tiny sample of those eligible. Certainly the laws that provide for this punishment do not attempt to draw that distinction; all cases to which the laws apply are necessarily "extreme." Nor is the distinction credible in fact. If, for example, petitioner Furman or his crime illustrates the "extreme," then nearly all murderers and their murders are also "extreme."[1] Furthermore, our procedures in death cases, rather than resulting in the selection of "extreme" cases for this punishment, actually sanction an arbitrary selection. For this Court has held that juries may, as they do, make the decision

1. The victim surprised Furman in the act of burglarizing the victim's home in the middle of the night. While escaping, Furman killed the victim with one pistol shot fired through the closed kitchen door from the outside. At the trial, Furman gave his version of the killing:

"They got me charged with murder and I admit, I admit going to these folks' home and they did caught me in there and I was coming back out, backing up and there was a wire down there on the floor. I was coming out backwards and fell back and I didn't intend to kill nobody. I didn't know they was behind the door. The gun went off and I didn't know nothing about no murder until they arrested me, and when the gun went off I was down on the floor and I got up and ran. That's all to it."

The Georgia Supreme Court accepted that version:

"The admission in open court by the accused . . . that during the period in which he was involved in the commission of a criminal act at the home of the deceased, he accidentally tripped over a wire in leaving the premises causing the gun to go off, together with other facts and circumstances surrounding the death of the deceased by violent means, was sufficient to support the verdict of guilty of murder. . . ."

About Furman himself, the jury knew only that he was black and that according to his statement at trial, he was 26 years old and worked at "Superior Upholstery." It took the jury one hour and 35 minutes to return a verdict of guilt and a sentence of death.

whether to impose a death sentence wholly unguided by standards governing that decision. In other words, our procedures are not constructed to guard against the totally capricious selection of criminals for the punishment of death.

Although it is difficult to imagine what further facts would be   13
necessary in order to prove that death is, as my Brother [Potter] Stewart puts it, "wantonly and . . . freakishly" inflicted, I need not conclude that arbitrary infliction is patently obvious. I am not considering this punishment by the isolated light of one principle. The probability of arbitrariness is sufficiently substantial that it can be relied upon, in combination with the other principles, in reaching a judgment on the constitutionality of this punishment.

When there is a strong probability that an unusually severe and   14
degrading punishment is being inflicted arbitrarily, we may well expect that society will disapprove of its infliction. I turn, therefore, to the third principle. An examination of the history and present operation of the American practice of punishing criminals by death reveals that this punishment has been almost totally rejected by contemporary society.

I cannot add to my Brother [Thurgood] Marshall's comprehen-   15
sive treatment of the English and American history of this punish-ment. I emphasize, however, one significant conclusion that emerges from that history. From the beginning of our Nation, the punishment of death has stirred acute public controversy. Although pragmatic arguments for and against the punishment have been frequently advanced, this longstanding and heated controversy cannot be explained solely as the result of differences over the practical wisdom of a particular government policy. At bottom, the battle has been waged on moral grounds. The country has debated whether a society for which the dignity of the individual is the supreme value can, without a fundamental inconsistency, follow the practice of deliberately putting some of its members to death. In the United States, as in other nations of the western world, "the struggle about this punishment has been one between ancient and deeply rooted beliefs in retribution, atonement or vengeance on the one hand, and, on the other, beliefs in the personal value and dignity of the common man that were born of the democratic movement of the eighteenth century, as well as beliefs in the scientific approach

to an understanding of the motive forces of human conduct, which are the result of the growth of the sciences of behavior during the nineteenth and twentieth centuries." It is this essentially moral conflict that forms the backdrop for the past changes in and the present operation of our system of imposing death as a punishment for crime.

16   Our practice of punishing criminals by death has changed greatly over the years. One significant change has been in our methods of inflicting death. Although this country never embraced the more violent and repulsive methods employed in England, we did for a long time rely almost exclusively upon the gallows and the firing squad. Since the development of the supposedly more humane methods of electrocution late in the 19th century and lethal gas in the 20th, however, hanging and shooting have virtually ceased. Our concern for decency and human dignity, moreover, has compelled changes in the circumstances surrounding the execution itself. No longer does our society countenance the spectacle of public executions, once thought desirable as a deterrent to criminal behavior by others. Today we reject public executions as debasing and brutalizing to us all.

17   Also significant is the drastic decrease in the crimes for which the punishment of death is actually inflicted. While esoteric capital crimes remain on the books, since 1930 murder and rape have accounted for nearly 99% of the total executions, and murder alone for about 87%. In addition, the crime of capital murder has itself been limited. As the Court noted in *McGautha* v. *California*, there was in this country a "rebellion against the common-law rule imposing a mandatory death sentence on all convicted murderers." Initially, that rebellion resulted in legislative definitions that distinguished between degrees of murder, retaining the mandatory death sentence only for murder in the first degree. Yet "[t]his new legislative criterion for isolating crimes appropriately by death soon proved as unsuccessful as the concept of 'malice aforethought,'" the common-law means of separating murder from manslaughter. Not only was the distinction between degrees of murder confusing and uncertain in practice, but even in clear cases of first-degree murder juries continued to take the law into their own hands: if they felt that death was an inappropriate punishment, "they simply refused

to convict of the capital offense." The phenomenon of jury nullification thus remained to counteract the rigors of mandatory death sentences. Bowing to reality, "legislatures did not try, as before, to refine further the definition of capital homicides. Instead they adopted the method of forthrightly granting juries the discretion which they had been exercising in fact." In consequence, virtually all death sentences today are discretionarily imposed. Finally, it is significant that nine States no longer inflict the punishment of death under any circumstances, and five others have restricted it to extremely rare crimes.

Thus, although "the death penalty has been employed throughout our history," in fact the history of this punishment is one of successive restriction. What was once a common punishment has become, in the context of a continuing moral debate, increasingly rare. The evolution of this punishment evidences, not that it is an inevitable part of the American scene, but that it has proved progressively more troublesome to the national conscience. The result of this movement is our current system of administering the punishment, under which death sentences are rarely imposed and death is even more rarely inflicted. It is, of course, "We, the People" who are responsible for the rarity both of the imposition and the carrying out of this punishment. Juries, "express[ing] the conscience of the community on the ultimate question of life or death," *Witherspoon* v. *Illinois,* have been able to bring themselves to vote for death in a mere 100 or so cases among the thousands tried each year where the punishment is available. Governors, elected by and acting for us, have regularly commuted a substantial number of those sentences. And it is our society that insists upon due process of law to the end that no person will be unjustly put to death, thus ensuring that many more of those sentences will not be carried out. In sum, we have made death a rare punishment today.

The progressive decline in, and the current rarity of, the infliction of death demonstrate that our society seriously questions the appropriateness of this punishment today. The States point out that many legislatures authorize death as the punishment for certain crimes and that substantial segments of the public, as reflected in opinion polls and referendum votes, continue to support it. Yet the availability of this punishment through statutory authorization, as well

18

19

as the polls and referenda, which amount simply to approval of that authorization, simply underscores the extent to which our society has in fact rejected this punishment. When an unusually severe punishment is authorized for wide-scale application but not, because of society's refusal, inflicted save in a few instances, the inference is compelling that there is a deep-seated reluctance to inflict it. Indeed, the likelihood is great that the punishment is tolerated only because of its disuse. The objective indicator of society's view of an unusually severe punishment is what society does with it, and today society will inflict death upon only a small sample of the eligible criminals. Rejection could hardly be more complete without becoming absolute. At the very least, I must conclude that contemporary society views this punishment with substantial doubt.

20   The final principle to be considered is that an unusually severe and degrading punishment may not be excessive in view of the purposes for which it is inflicted. This principle, too, is related to the others. When there is a strong probability that the State is arbitrarily inflicting an unusually severe punishment that is subject to grave societal doubts, it is likely also that the punishment cannot be shown to be serving any penal purpose that could not be served equally well by some less severe punishment.

21   The States' primary claim is that death is a necessary punishment because it prevents the commission of capital crimes more effectively than any less severe punishment. The first part of this claim is that the infliction of death is necessary to stop the individuals executed from committing further crimes. The sufficient answer to this is that if a criminal convicted of a capital crime poses a danger to society, effective administration of the State's pardon and parole laws can delay or deny his release from prison, and techniques of isolation can eliminate or minimize the danger while he remains confined.

22   The more significant argument is that the threat of death prevents the commission of capital crimes because it deters potential criminals who would not be deterred by the threat of imprisonment. The argument is not based upon evidence that the threat of death is a superior deterrent. Indeed, as my Brother Marshall establishes, the available evidence uniformly indicates, although it does not conclusively prove, that the threat of death has no greater deterrent

effect than the threat of imprisonment. The States argue, however, that they are entitled to rely upon common human experience, and that experience, they say, supports the conclusion that death must be a more effective deterrent than any less severe punishment. Because people fear death the most, the argument runs, the threat of death must be the greatest deterrent.

It is important to focus upon the precise import of this argument. 23
It is not denied that many, and probably most, capital crimes cannot be deterred by the threat of punishment. Thus the argument can apply only to those who think rationally about the commission of capital crimes. Particularly is that true when the potential criminal, under this argument, must not only consider the risk of punishment, but also distinguish between two possible punishments. The concern, then, is with a particular type of potential criminal, the rational person who will commit a capital crime knowing that the punishment is long-term imprisonment, which may well be for the rest of his life, but will not commit the crime knowing that the punishment is death. On the face of it, the assumption that such persons exist is implausible.

In any event, this argument cannot be appraised in the abstract. 24
We are not presented with the theoretical question whether under any imaginable circumstances the threat of death might be a greater deterrent to the commission of capital crimes than the threat of imprisonment. We are concerned with the practice of punishing criminals by death as it exists in the United States today. Proponents of this argument necessarily admit that its validity depends upon the existence of a system in which the punishment of death is invariably and swiftly imposed. Our system, of course, satisfies neither condition. A rational person contemplating a murder or rape is confronted, not with the certainty of a speedy death, but with the slightest possibility that he will be executed in the distant future. The risk of death is remote and improbable; in contrast, the risk of long-term imprisonment is near and great. In short, whatever the speculative validity of the assumption that the threat of death is a superior deterrent, there is no reason to believe that as currently administered the punishment of death is necessary to deter the commission of capital crimes. Whatever might be the case were all or substantially all eligible criminals quickly put to death, unverifiable possibilities are an insufficient basis upon which to conclude

that the threat of death today has any greater deterrent efficacy than the threat of imprisonment.[2]

25   There is, however, another aspect to the argument that the punishment of death is necessary for the protection of society. The infliction of death, the States urge, serves to manifest the community's outrage at the commission of the crime. It is, they say, a concrete public expression of moral indignation that inculcates respect for the law and helps assure a more peaceful community. Moreover, we are told, not only does the punishment of death exert this widespread moralizing influence upon community values, it also satisfies the popular demand for grievous condemnation of abhorrent crimes and thus prevents disorder, lynching, and attempts by private citizens to take the law into their own hands.

26   The question, however, is not whether death serves these supposed purposes of punishment, but whether death serves them more effectively than imprisonment. There is no evidence whatever that utilization of imprisonment rather than death encourages private blood feuds and other disorders. Surely if there were such a danger, the execution of a handful of criminals each year would not prevent it. The assertion that death alone is a sufficiently emphatic denunciation for capital crimes suffers from the same defect. If capital crimes require the punishment of death in order to provide moral reinforcement for the basic values of the community, those values can only be undermined when death is so rarely inflicted upon the criminals who commit the crimes. Furthermore, it is certainly doubtful that the infliction of death by the State does in fact strengthen the community's moral code; if the deliberate extinguishment of human life has any effect at all, it more likely tends to lower our respect for life and brutalize our values. That, after all,

2. There is also the more limited argument that death is a necessary punishment when criminals are already serving or subject to a sentence of life imprisonment. If the only punishment available is further imprisonment, it is said, those criminals will have nothing to lose by committing further crimes, and accordingly the threat of death is the sole deterrent. But "life" imprisonment is a misnomer today. Rarely, if ever, do crimes carry a mandatory life sentence without possibility of parole. That possibility ensures that criminals do not reach the point where further crimes are free of consequences. Moreover, if this argument is simply an assertion that the threat of death is a more effective deterrent than the threat of increased imprisonment by denial of release on parole, then, as noted above, there is simply no evidence to support it.

is why we no longer carry out public executions. In any event, this claim simply means that one purpose of punishment is to indicate social disapproval of crime. To serve that purpose our laws distribute punishments according to the gravity of crimes and punish more severely the crimes society regards as more serious. That purpose cannot justify any particular punishment as the upper limit of severity.

There is, then, no substantial reason to believe that the punishment of death, as currently administered, is necessary for the protection of society. The only other purpose suggested, one that is independent of protection for society, is retribution. Shortly stated, retribution in this context means that criminals are put to death because they deserve it.

Although it is difficult to believe that any State today wishes to proclaim adherence to "naked vengeance," the States claim, in reliance upon its statutory authorization, that death is the only fit punishment for capital crimes and that this retributive purpose justifies its infliction. In the past, judged by its statutory authorization, death was considered the only fit punishment for the crime of forgery, for the first federal criminal statute provided a mandatory death penalty for that crime. Obviously, concepts of justice change; no immutable moral order requires death for murderers and rapists. The claim that death is a just punishment necessarily refers to the existence of certain public beliefs. The claim must be that for capital crimes death alone comports with society's notion of proper punishment. As administered today, however, the punishment of death cannot be justified as a necessary means of exacting retribution from criminals. When the overwhelming number of criminals who commit capital crimes go to prison, it cannot be concluded that death serves the purpose of retribution more effectively than imprisonment. The asserted public belief that murderers and rapists deserve to die is flatly inconsistent with the execution of a random few. As the history of the punishment of death in this country shows, our society wishes to prevent crime; we have no desire to kill criminals simply to get even with them.

In sum, the punishment of death is inconsistent with all four principles: Death is an unusually severe and degrading punishment; there is a strong probability that it is inflicted arbitrarily; its rejection by contemporary society is virtually total; and there is no reason to believe that it serves any penal purpose more effectively than the less severe punishment of imprisonment. The function of these principles is to enable a court to determine whether a punishment comports with human dignity. Death, quite simply, does not. . . .

STUDY QUESTIONS

1. In paragraphs 7 to 14, Brennan argues that the death penalty is being inflicted arbitrarily. The following propositions are the main premises and conclusions of his argument. Diagram the argument.
   (1) The use of the death penalty has declined since the 1930s (paragraph 8).
   (2) The death penalty is not regularly and fairly applied (paragraph 9).
   (3) The death penalty is not used in all cases where authorized as a punishment (paragraph 10).
   (4) The use of the death penalty is rare (paragraph 10).
   (5) The death penalty is being arbitrarily inflicted (paragraph 11).
   (6) The rarity of the death penalty is not because of informed selectivity (paragraph 12).
   (7) Crimes and criminals do not admit of a distinction that can be drawn so finely as to explain the execution of such a tiny sample of those eligible (paragraph 12).
   (8) Assumed premise: If the rarity of the death penalty were because of informed selectivity, then we should be able to draw a fine distinction between the extreme crimes and criminals for which the death penalty is used and the less extreme crimes and criminals for which the death penalty is not used.
2. Using question (1) as a model, diagram Brennan's arguments for the conclusions that the death penalty is unusually degrading to human dignity (paragraphs 2 to 6); that society disapproves of the infliction of the death penalty (paragraphs 14 to 19); and that the use of the death penalty is an excessive punishment (paragraphs 20–28).

3. Many people believe that the death penalty is justified because it serves as a deterrent for potential criminals while also ensuring that convicted criminals will not commit additional crimes. In paragraph 27, Brennan concludes that there is "no substantial reason to believe that the punishment of death, as currently administered, is necessary for the protection of society." Should the burden of proof be on Brennan to prove that society will be just as safe if the death penalty is not used, or should it be on Brennan's opponents to prove that the death penalty makes society safer?

Reference Chapters: 6, 7

PART TWO

On Adultery and Murder

GORGIAS OF LEONTINOI

Encomium of Helen

Gorgias (c. 483–376 B.C.) was a native of Leontinoi in Sicily. He went to Athens in 427 B.C. as ambassador from Leontinoi. There he became an influential orator and teacher of rhetoric: his students included the famous statesman Pericles and the historian Thucydides. The following was probably a ceremonial speech, one of the few of Gorgias' works that have survived. In it, Gorgias takes on the challenge of defending the beautiful Helen of Troy against the charge of adultery. As Greek legend has it, Helen sparked the beginning of the Trojan War when she betrayed her husband Menelaus, the king

[Gorgias, "Encomium of Helen," transl. Douglas Maurice MacDowell. Bristol, Engl.: Bristol Classical Books, 1982, pp. 21, 23, 25, 27, 29, 31.]

of Sparta, by running away with Alexander, the prince of Troy.

1   The grace of a city is excellence of its men, of a body beauty, of a mind wisdom, of an action virtue, of a speech truth; the opposites of these are a disgrace. A man, a woman, a speech, a deed, a city, and an action, if deserving praise, one should honour with praise, but to the undeserving one should attach blame. For it is an equal error and ignorance to blame the praiseworthy and to praise the blameworthy. The man who says rightly what ought to be said should also refute those who blame Helen, a woman about whom both the belief of those who have listened to poets and the message of her name, which has become a reminder of the calamities, have been in unison and unanimity. I wish, by adding some reasoning to my speech, to free the slandered woman from the accusation and to demonstrate that those who blame her are lying, and both to show what is true and to put a stop to their ignorance.

2   That the woman who is the subject of this speech was pre-eminent among pre-eminent men and women, by birth and descent, is not obscure to even a few. It is clear that her mother was Leda, and her actual father was a god and her reputed father a mortal, Tyndareos and Zeus, of whom the one was believed to be because he was and the other was reputed to be because he said he was, and the one was the best of men and the other the master of all. Born of such parents, she had godlike beauty, which she acquired and had openly. In very many she created very strong amorous desires; with a single body she brought together many bodies of men who had great pride for great reasons; some of them had great amounts of wealth, others fame of ancient nobility, others vigour of personal strength, others power of acquired wisdom; and they all came because of a love which wished to conquer and a wish for honour which was unconquered. Who fulfilled his love by obtaining Helen, and why, and how, I shall not say; for to tell those who know what they know carries conviction but does not give pleasure. Passing over now in my speech that former time, I shall proceed to the beginning of my intended speech, and I shall propound the causes which made it reasonable for Helen's departure to Troy to occur.

3   Either it was because of the wishes of Chance and the purposes of the gods and the decrees of Necessity that she did what she did, or because she was seized by force, or persuaded by speeches, (or

captivated by love). Now, if it was because of the first, the accuser deserves to be accused; for it is impossible to hinder a god's predetermination by human preconsideration. It is not natural for the stronger to be hindered by the weaker, but for the weaker to be governed and guided by the stronger, and for the stronger to lead and the weaker to follow. A god is a stronger thing than a human being, both in force and in wisdom and in other respects. So if the responsibility is to be attributed to Chance and God, Helen is to be released from the infamy.

But if she was seized by force and unlawfully violated and 4 unjustly assaulted, clearly the man who seized or assaulted did wrong, and the woman who was seized or was assaulted suffered misfortune. So the barbarian who undertook a barbaric undertaking in speech and in law and in deed deserves to receive accusation in speech, debarment in law, and punishment in deed; but the woman who was violated and deprived of her country and bereaved of her family, would she not reasonably be pitied rather than reviled? He performed terrible acts, she suffered them; so it is just to sympathize with her but to hate him.

But if it was speech that persuaded and deceived her mind, it is 5 also not difficult to make a defence for that and to dispel the accusation thus. Speech is a powerful ruler. Its substance is minute and invisible, but its achievements are superhuman; for it is able to stop fear and remove sorrow and to create joy and to augment pity. I shall prove that this is so; I must also prove it by opinion to my hearers.

All poetry I consider and call speech with metre. Into those who 6 hear it comes fearful fright and tearful pity and mournful longing, and at the successes and failures of others' affairs and persons the mind suffers, through speeches, a suffering of its own.

Now then, let me move from one speech to another. Inspired 7 incantations through speeches are inducers of pleasure and reducers of sorrow; by intercourse with the mind's belief, the power of the incantation enchants and persuades and moves it by sorcery. Two arts of sorcery and magic have been invented; they are deviations of mind and deceptions of belief.

How many men have persuaded and do persuade how many, on 8 how many subjects, by fabricating false speech! For if everyone, on every subject, possessed memory of the past and (understanding) of the present and foreknowledge of the future, speech would not be equally (powerful); but as it is, neither remembering a past event nor investigating a present one nor prophesying a future one is easy, so that on most subjects most men make belief their mind's adviser. But belief, being slippery and unreliable, brings slippery and unreliable success to those who employ it. So what reason is there against Helen's also (having come under the influence of speech just as much against her will as if she had been seized by violence of violators? For persuasion expelled sense; and indeed persuasion, though not having an appearance of compulsion,) has the same power. For speech, the persuader, compelled mind, the persuaded, both to obey what was said and to approve what was done. So the persuader, because he compelled, is guilty; but the persuaded, because she was compelled by his speech, is wrongly reproached.

9   To show that persuasion, when added to speech, also moulds the mind in the way it wishes, one should note first the speeches of astronomers, who substituting belief for belief, demolishing one and establishing another, make the incredible and obscure become clear to the eyes of belief; and secondly compulsory contests conducted by means of speeches, in which a single speech pleases and persuades a large crowd, because written with skill, not spoken with truth; (and) thirdly conflicts of philosophical speeches, in which it is shown that quick-wittedness too makes the opinion which is based on belief changeable. The power of speech bears the same relation to the ordering of the mind as the ordering of drugs bears to the constitution of bodies. Just as different drugs expel different humours from the body, and some stop it from being ill but others stop it from living, so too some speeches cause sorrow, some cause pleasure, some cause fear, some give the hearers confidence, some drug and bewitch the mind with an evil persuasion.

10   That, if she was persuaded by speech, it was not a misdeed but a mischance, has been stated; and I shall examine the fourth cause in the fourth part of my speech. If it was love that brought all this about, she will without difficulty escape the accusation of the offence said to have been committed. Things that we see do not have the nature which we wish them to have but the nature which each of them actually has; and by seeing them the mind is moulded in its character too. For instance, when the sight surveys hostile persons and a hostile array of bronze and iron for hostile armament, offensive array of the one and shields of the other, it is alarmed, and it alarms the mind, so that often people flee in panic when some

danger is imminent as if it were present. So strong is the disregard of law which is implanted in them because of the fear caused by the sight; when it befalls, it makes them disregard both the honour which is awarded for obeying the law and the benefit which accrues for doing right. And some people before now, on seeing frightful things, have also lost their presence of mind at the present moment; fear so extinguishes and expels thought. And many have fallen into groundless distress and terrible illness and incurable madness; so deeply does sight engrave on the mind images of actions that are seen. And as far as frightening things are concerned, many are omitted, but those omitted are similar to those mentioned.

But when painters complete out of many colours and objects a single object and form, they please the sight. The making of figures and the creation of statues provides a pleasant disease for the eyes. Thus some things naturally give distress and others pleasure to the sight. Many things create in many people love and desire of many actions and bodies. So if Helen's eye, pleased by Alexander's body, transmitted an eagerness and striving of love to her mind, what is surprising? If love is a god with a god's power, how would the weaker be able to repel and resist it? But if it is a human malady and incapacity of mind, it should not be blamed as an impropriety but considered as an adversity; for it comes, when it does come, through deceptions of mind, not intentions of thought, and through compulsions of love, not contrivances of skill. 11

So how should one consider the blame of Helen just? Whether she did what she did because she was enamoured (by sight) or persuaded by speech or seized by force or compelled by divine necessity, in every case she escapes the accusation. 12

I have removed by my speech a woman's infamy, I have kept to the purpose which I set myself at the start of my speech; I attempted to dispel injustice of blame and ignorance of belief, I wished to write the speech as an encomium of Helen and an amusement for myself. 13

STUDY QUESTIONS

1. How many possible explanations for Helen's actions does Gorgias consider? Are these the only possibilities?

2. According to Gorgias' argument, under what conditions can one properly be blamed?

3. What powers does Gorgias have to ascribe to speech and love in order to absolve Helen?

4. Notice that in paragraph 8, Gorgias says, "How many men have persuaded and do persuade how many, on how many subjects, by fabricating false speech!" What does this imply for his argument in defense of Helen? How can we be sure that Gorgias isn't misleading us by false speech?

Reference Chapters: 10, 11

LYSIAS

On the Murder of Eratosthenes: Defence

Lysias (c. 450–c. 380 B.C.) was an orator, teacher of rhetoric, and professional speechwriter. For his impassioned defenses of democracy, he was banished from Thurii, the city he grew up in. He then lived in Athens, but was forced into exile by the Thirty, a group of tyrants who ruled Athens for a short time following the end of the Peloponnesian War in 404 B.C. After the fall of the Thirty, Lysias returned to Athens but again lost the rights of a citizen and so was unable to speak in the assembly or in court. He made a living as a writer of courtroom speeches. The following selection is one of eight hundred that

[Lysias, "On the Murder of Eratosthenes," transl. W. R. M. Lamb. Cambridge, Mass.: Harvard University Press, 1957, odd-numbered pp. 5–27, some footnotes omitted.]

*Lysias wrote, of which only twenty-three survive. It was writ-
ten for Euphiletus, who had killed a member of the Thirty,
Eratosthenes, who had been sleeping with Euphiletus' wife.
Unfortunately, we do not know the verdict in the case.*

Defence

I should be only too pleased, sirs, to have you so disposed towards   1
me in judging this case as you would be to yourselves, if you found
yourselves in my plight. For I am sure that, if you had the same
feelings about others as about yourselves, not one of you but would
be indignant at what has been done; you would all regard the
penalties appointed for those who resort to such practices as too
mild. And these feelings would be found, not only amongst you, but
in the whole of Greece: for in the case of this crime alone, under
both democracy and oligarchy, the same requital is accorded to the
weakest against the strongest, so that the lowest gets the same
treatment as the highest. Thus you see, sirs, how all men abominate
this outrage. Well, I conceive that, in regard to the severity of the
penalty, you are all of the same mind, and that not one of you is so
easygoing as to think it right that men who are guilty of such acts
should obtain pardon, or to presume that slight penalties suffice for
their deserts. But I take it, sirs, that what I have to show is that
Eratosthenes had an intrigue with my wife, and not only corrupted
her but inflicted disgrace upon my children and an outrage on
myself by entering my house; that this was the one and only enmity
between him and me; that I have not acted thus for the sake of
money, so as to raise myself from poverty to wealth; and that all I
seek to gain is the requital accorded by our laws. I shall therefore
set forth to you the whole of my story from the beginning; I shall
omit nothing, but will tell the truth. For I consider that my own sole
deliverance rests on my telling you, if I am able, the whole of what
has occurred.

When I, Athenians, decided to marry, and brought a wife into   2
my house, for some time I was disposed neither to vex her nor to
leave her too free to do just as she pleased; I kept a watch on her as
far as possible, with such observation of her as was reasonable. But

when a child was born to me, thenceforward I began to trust her, and placed all my affairs in her hands, presuming that we were now in perfect intimacy. It is true that in the early days, Athenians, she was the most excellent of wives; she was a clever, frugal house-keeper, and kept everything in the nicest order. But as soon as I lost my mother, her death became the cause of all my troubles. For it was in attending her funeral that my wife was seen by this man, who in time corrupted her. He looked out for the servant-girl who went to market, and so paid addresses to her mistress by which he wrought her ruin. Now in the first place I must tell you, sirs (for I am obliged to give you these particulars), my dwelling is on two floors, the upper being equal in space to the lower, with the women's quarters above and the men's below. When the child was born to us, its mother suckled it; and in order that, each time that it had to be washed, she might avoid the risk of descending by the stairs, I used to live above, and the women below. By this time it had become such an habitual thing that my wife would often leave me and go down to sleep with the child, so as to be able to give it the breast and stop its crying. Things went on in this way for a long time, and I never suspected, but was simpleminded enough to suppose that my own was the chastest wife in the city. Time went on, sirs; I came home unexpectedly from the country, and after dinner the child started crying in a peevish way, as the servant-girl was annoying it on purpose to make it so behave; for the man was in the house,—I learnt it all later. So I bade my wife go and give the child her breast, to stop its howling. At first she refused, as though delighted to see me home again after so long; but when I began to be angry and bade her go,—"Yes, so that you," she said, "may have a try here at the little maid. Once before, too, when you were drunk, you pulled her about." At that I laughed, while she got up, went out of the room, and closed the door, feigning to make fun, and she turned the key in the lock. I, without giving a thought to the matter, or having any suspicion, went to sleep in all content after my return from the country. Towards daytime she came and opened the door. I asked why the doors made a noise in the night; she told me that the child's lamp had gone out, and she had lit it again at our neighbour's. I was silent and believed it was so. But it struck me, sirs, that she had powdered her face, though her brother had died not thirty days before; even so, however, I made no remark on the fact, but left the house in silence. After this, sirs, an interval occurred in which I was left quite unaware of my own injuries; I was then accosted by a certain old female, who was secretly sent by a woman with whom that man was having an intrigue, as I heard later. This woman was angry with him and felt herself wronged, because he no longer visited her so regularly, and she kept a close watch on him until she discovered what was the cause. So the old creature accosted me where she was on the look-out, near my house, and said,— "Euphiletus, do not think it is from any meddlesomeness that I have approached you; for the man who is working both your and your wife's dishonour happens to be our enemy. If, therefore, you take the servant-girl who goes to market and waits on you, and torture her, you will learn all. It is," she said, "Eratosthenes of Oë who is doing this; he has debauched not only your wife, but many others besides; he makes an art of it." With these words, sirs, she took herself off; I was at once perturbed; all that had happened came into my mind, and I was filled with suspicion,—reflecting first how I was shut up in my chamber, and then remembering how on that night the inner and outer doors made a noise, which had never occurred before, and how it struck me that my wife had put on powder. All these things came into my mind, and I was filled with suspicion. Returning home, I bade the servant-girl follow me to the market, and taking her to the house of an intimate friend, I told her I was fully informed of what was going on in my house: "So it is open to you," I said, "to choose as you please between two things,— either to be whipped and thrown into a mill, never to have any rest from miseries of that sort, or else to speak out the whole truth and, instead of suffering any harm, obtain my pardon for your transgressions. Tell no lies, but speak the whole truth." The girl at first denied it, and bade me do what I pleased, for she knew nothing; but when I mentioned Eratosthenes to her, and said that he was the man who visited my wife, she was dismayed, supposing that I had exact knowledge of everything. At once she threw herself down at my knees, and having got my pledge that she should suffer no harm, she accused him, first, of approaching her after the funeral, and then told how at last she became his messenger; how my wife in time was persuaded, and by what means she procured his entrances, and how at the Thesmophoria, while I was in the country, she went off to the temple with his mother. And the girl gave an exact account of everything else that had occurred. When her tale was all told, I said,—"Well now, see that nobody in the world gets knowledge of this; otherwise, nothing in your arrangement with me will hold good. And I require that you show me their guilt in the very act; I want no words, but manifestation of the fact, if it really is so." She agreed to do this. Then came an interval of four or five days . . . as I shall bring strong evidence to show. But first I wish to relate what took place on the last day. I had an intimate friend named Sostratus. After sunset I met him as he came from the country. As I knew that, arriving at that hour, he would find none of his circle at home, I invited him to dine with me; we came to my house, mounted to the upper room, and had dinner. When he had made a good meal, he left me and departed; then I went to bed. Eratosthenes, sirs, entered, and the maid-servant roused me at once, and told me that he was in the house. Bidding her look after the door, I descended and went out in silence; I called on one friend and another, and found some of them at home, while others were out of town. I took with me as many as I could among those who were there, and so came along. Then we got torches from the nearest shop, and went in; the door was open, as the girl had it in readiness. We pushed open the door of the bedroom, and the first of us to enter were in time to see him lying down by my wife; those who followed saw him standing naked on the bed. I gave him a blow, sirs, which knocked him down, and pulling round his two hands behind his back, and tying them, I asked him why he had the insolence to enter my house. He admitted his guilt; then he besought and implored me not to kill him, but to exact a sum of money. To this I replied,—"It is not I who am going to kill you, but our city's law, which you have transgressed and regarded as of less account than your pleasures, choosing rather to commit this foul offence against my wife and my children than to obey the laws like a decent person."

2   Thus it was, sirs, that this man incurred the fate that the laws ordain for those who do such things;[1] he had not been dragged in there from the street, nor had he taken refuge at my hearth,[2] as these people say. For how could it be so, when it was in the bedroom that

1. Athenian law provided that a husband could kill his wife's seducer if the killing was unpremeditated. [Editor's note.]
2. The hearth in a Greek house retained its primitive sanctity as a centre of the family religion, and it would be sacrilege to kill anyone there. [Translator's note.]

he was struck and fell down then and there, and I pinioned his arms, and so many persons were in the house that he could not escape them, as he had neither steel nor wood nor anything else with which he might have beaten off those who had entered? But, sirs, I think you know as well as I that those whose acts are against justice do not acknowledge that their enemies speak the truth, but lie themselves and use other such devices to foment anger in their hearers against those whose acts are just. So, first read the law.

Law [The law is read]

He did not dispute it, sirs: he acknowledged his guilt, and besought and implored that he might not be killed, and was ready to pay compensation in money. But I would not agree to his estimate, as I held that our city's law should have higher authority; and I obtained that satisfaction which you deemed most just when you imposed it on those who adopt such courses. Now, let my witnesses come forward in support of these statements. 4

Witnesses [Witnesses are heard]

Read out also, please, that law from the pillar in the Areopagus. 5

Law [The law is read]

You hear, sirs, how the Court of the Areopagus itself, to which has been assigned, in our own as in our fathers' time, the trial of suits for murder, has expressly stated that whoever takes this vengeance on an adulterer caught in the act with his spouse shall not be convicted of murder. And so strongly was the lawgiver convinced of the justice of this in the case of wedded wives, that he even applied the same penalty in the case of mistresses, who are of less account. Now surely it is clear that, if he had had any heavier punishment than this for the case of married women, he would have imposed it. 6

But in fact, as he was unable to devise a severer one for wives, he ordained that it should be the same for that of mistresses also. Please read this law besides.

Law [The law is read]

7   You hear, sirs, how it directs that, if anyone forcibly debauches a free adult or child, he shall be liable to double damages; while if he so debauches a woman, in any of the cases where it is permitted to kill him, he is subject to the same rule. Thus the lawgiver, sirs, considered that those who use force deserve a less penalty than those who use persuasion; for the latter he condemned to death, whereas for the former he doubled the damages, considering that those who achieve their ends by force are hated by the persons forced; while those who used persuasion corrupted thereby their victims' souls, thus making the wives of others more closely attached to themselves than to their husbands, and got the whole house into their hands, and caused uncertainty as to whose the children really were, the husbands' or the adulterers'. In view of all this the author of the law made death their penalty. Wherefore I, sirs, not only stand acquitted of wrongdoing by the laws, but am also directed by them to take this satisfaction: it is for you to decide whether they are to be valid or of no account. For to my thinking every city makes its laws in order that on any matter which perplexes us we may resort to them and inquire what we have to do. And so it is they who, in cases like the present, exhort the wronged parties to obtain this kind of satisfaction. I call upon you to support their opinion: otherwise, you will be giving adulterers such license that you will encourage thieves as well to call themselves adulterers; since they will feel assured that, if they plead this reason in their defence, and allege that they enter other men's houses for this purpose, nobody will touch them. For everyone will know that the laws on adultery are to be dismissed, and that it is your vote that one has to fear, because this has supreme authority over all the city's affairs.

8   Do not consider, sirs, what they say: they accuse me of ordering the maid-servant on that day to go and fetch the young man. Now I, sirs, could have held myself justified in using any possible means to catch the corrupter of my wife. For if I had bidden the girl fetch

him, when words alone had been spoken and no act had been committed, I should have been in the wrong: but if, when once he had compassed all his ends, and had frequently entered my house, I had then used any possible means to catch him, I should have considered myself quite in order. And observe how on this point also they are lying: you will perceive it easily in this way. As I told you, sirs, before, Sostratus was a friend of mine, on intimate terms with me; he met me as he came from the country about sunset, and had dinner with me, and when he had made a good meal he left me and departed. Now in the first place, sirs, you must bear this in mind: if on that night I had designs on Eratosthenes, which was more to my advantage,—to go and take my dinner elsewhere, or to bring in my guest to dinner with me? For in the latter case that man would have been less likely to venture on entering my house. And in the second place, do you suppose that I should have let my dinner-guest go and leave me there alone and unsupported, and not rather have bidden him stay, in order that he might stand by me in taking vengeance upon the adulterer? Then again, sirs, do you not think that I should have sent word to my intimate acquaintances in the daytime, and bidden them assemble at the house of one of my friends living nearest to me, rather than have waited till the moment of making my discovery to run round in the night, without knowing whom I should find at home, and who were away? Thus I called on Harmodius, and one other, who were not in town—of this I was not aware—and others, I found, were not in; but those whom I could I took along with me. Yet if I had foreknown this, do you not think that I should have called up servants and passed the word to my friends, in order that I might have gone in myself with all possible safety,—for how could I tell whether he too had some weapon?— and so I might have had as many witnesses as possible with me when I took my vengeance? But as in fact I knew nothing of what was to befall on that night, I took with me those whom I could. Now let my witnesses come forward in support of all this.

Witnesses [Witnesses are heard]

You have heard the witnesses, sirs; and consider this affair thus in your own minds, asking yourselves whether any enmity has ever

arisen before this between me and Eratosthenes. I say you will
discover none. For he had neither subjected me to slanderous
impeachment, nor attempted to expel me from the city, nor brought
any private suit against me, nor was he privy to any wrongdoing
which I was so afraid of being divulged that I was intent on his
destruction, nor, should I accomplish this, had I any hope of getting
money from anywhere: for there are people who plot each other's
death for such purposes. So far, indeed, from either abuse or a
drunken brawl or any other quarrel having occurred between us, I
had never even seen the man before that night. For what object,
then, should I run so grave a risk, unless I had received from him
the greatest of injuries? Why, again, did I choose to summon
witnesses for my wicked act, when it was open to me, if I was thus
criminally intent on his destruction, to have none of them privy to
it?

10   I therefore, sirs, do not regard this requital as having been exacted
in my own private interest, but in that of the whole city. For those
who behave in that way, when they see the sort of prizes offered for
such transgressions, will be less inclined to trespass against their
neighbours, if they see that you also take the same view. Otherwise
it were better far to erase our established laws, and ordain others
which will inflict the penalties on men who keep watch on their
own wives, and will allow full immunity to those who would
debauch them. This would be a far juster way than to let the citizens
be entrapped by the laws; these may bid a man, on catching an
adulterer, to deal with him in whatever way he pleases, but the trials
are found to be more dangerous to the wronged parties than to those
who, in defiance of the laws, dishonour the wives of others. For I
am now risking the loss of life, property and all else that I have,
because I obeyed the city's laws.

STUDY QUESTIONS

1. What do Lysias and Euphiletus consider to be justifiable
   circumstances for murder?
2. According to Athenian law, what distinct propositions must
   Euphiletus prove in order to show that he was justified in

killing Eratosthenes? Does Euphiletus address each of those propositions?

3. In paragraph 8, what strategy do Lysias and Euphiletus employ in responding to the charge that Eratosthenes was entrapped?

4. Was this a case of premeditated murder?

5. What does this selection tell us about Athenian customs in marriage? Does this information help to explain why Lysias doesn't discuss Euphiletus' wife's responsibility for the affair?

Reference Chapters: 7, 11

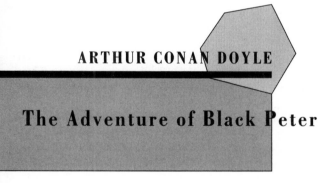

ARTHUR CONAN DOYLE

The Adventure of Black Peter

Sir Arthur Conan Doyle (1859–1930) was trained as a doctor, but is best known as the creator of the legendary detective Sherlock Holmes.

I have never known my friend to be in better form, both mental and physical, than in the year '95. His increasing fame had brought with it an immense practice, and I should be guilty of an indiscretion if I were even to hint at the identity of some of the illustrious clients who crossed our humble threshold in Baker Street. Holmes, how-

[Arthur Conan Doyle, "The Adventure of Black Peter," *The Return of Sherlock Holmes.* From Vol. II of *The Complete Sherlock Holmes.* New York: Doubleday & Co., pp. 558–572.]

ever, like all great artists, lived for his art's sake, and, save in the case of the Duke of Holdernesse, I have seldom known him claim any large reward for his inestimable services. So unworldly was he—or so capricious—that he frequently refused his help to the powerful and wealthy where the problem made no appeal to his sympathies, while he would devote weeks of most intense application to the affairs of some humble client whose case presented those strange and dramatic qualities which appealed to his imagination and challenged his ingenuity.

In this memorable year '95, a curious and incongruous succession of cases had engaged his attention, ranging from his famous investigation of the sudden death of Cardinal Tosca—an inquiry which was carried out by him at the express desire of His Holiness the Pope—down to his arrest of Wilson, the notorious canary-trainer, which removed a plague-spot from the East End of London. Close on the heels of these two famous cases came the tragedy of Woodman's Lee, and the very obscure circumstances which surrounded the death of Captain Peter Carey. No record of the doings of Mr. Sherlock Holmes would be complete which did not include some account of this very unusual affair.

During the first week of July, my friend had been absent so often and so long from our lodgings that I knew he had something on hand. The fact that several rough-looking men called during that time and inquired for Captain Basil made me understand that Holmes was working somewhere under one of the numerous disguises and names with which he concealed his own formidable identity. He had at least five small refuges in different parts of London, in which he was able to change his personality. He said nothing of his business to me, and it was not my habit to force a confidence. The first positive sign which he gave me of the direction which his investigation was taking was an extraordinary one. He had gone out before breakfast, and I had sat down to mine when he strode into the room, his hat upon his head and a huge barbed-headed spear tucked like an umbrella under his arm.

"Good gracious, Holmes!" I cried. "You don't mean to say that you have been walking about London with that thing?"

5   "I drove to the butcher's and back."

"The butcher's?"

"And I return with an excellent appetite. There can be no question, my dear Watson, of the value of exercise before breakfast. But I am prepared to bet that you will not guess the form that my exercise has taken."

"I will not attempt it."

He chuckled as he poured out the coffee.

"If you could have looked into Allardyce's back shop, you would have seen a dead pig swung from a hook in the ceiling, and a gentleman in his shirt sleeves furiously stabbing at it with this weapon. I was that energetic person, and I have satisfied myself that by no exertion of my strength can I transfix the pig with a single blow. Perhaps you would care to try?"

"Not for worlds. But why were you doing this?"

"Because it seemed to me to have an indirect bearing upon the mystery of Woodman's Lee. Ah, Hopkins, I got your wire last night, and I have been expecting you. Come and join us."

Our visitor was an exceedingly alert man, thirty years of age, dressed in a quiet tweed suit, but retaining the erect bearing of one who was accustomed to official uniform. I recognized him at once as Stanley Hopkins, a young police inspector, for whose future Holmes had high hopes, while he in turn professed the admiration and respect of a pupil for the scientific methods of the famous amateur. Hopkins's brow was clouded, and he sat down with an air of deep dejection.

"No, thank you, sir. I breakfasted before I came round. I spent the night in town, for I came up yesterday to report."

"And what had you to report?"

"Failure, sir, absolute failure."

"You have made no progress?"

"None."

"Dear me! I must have a look at the matter."

"I wish to heavens that you would, Mr. Holmes. It's my first big chance, and I am at my wit's end. For goodness' sake, come down and lend me a hand."

"Well, well, it just happens that I have already read all the available evidence, including the report of the inquest, with some care. By the way, what do you make of that tobacco pouch, found on the scene of the crime? Is there no clue there?"

Hopkins looked surprised.

"It was the man's own pouch, sir. His initials were inside it. And it was of sealskin—and he was an old sealer."

"But he had no pipe."

25   "No, sir, we could find no pipe. Indeed, he smoked very little, and yet he might have kept some tobacco for his friends."

"No doubt. I only mention it because, if I had been handling the case, I should have been inclined to make that the starting-point of my investigation. However, my friend, Dr. Watson, knows nothing of this matter, and I should be none the worse for hearing the sequence of events once more. Just give us some short sketches of the essentials."

Stanley Hopkins drew a slip of paper from his pocket.

"I have a few dates here which will give you the career of the dead man, Captain Peter Carey. He was born in '45—fifty years of age. He was a most daring and successful seal and whale fisher. In 1883 he commanded the steam sealer *Sea Unicorn*, of Dundee. He had then had several successful voyages in succession, and in the following year, 1884, he retired. After that he travelled for some years, and finally he bought a small place called Woodman's Lee, near Forest Row, in Sussex. There he has lived for six years, and there he died just a week ago to-day.

"There were some most singular points about the man. In ordinary life, he was a strict Puritan—a silent, gloomy fellow. His household consisted of his wife, his daughter, aged twenty, and two female servants. These last were continually changing, for it was never a very cheery situation, and sometimes it became past all bearing. The man was an intermittent drunkard, and when he had the fit on him he was a perfect fiend. He has been known to drive his wife and daughter out of doors in the middle of the night and flog them through the park until the whole village outside the gates was aroused by their screams.

30   "He was summoned once for a savage assault upon the old vicar, who had called upon him to remonstrate with him upon his conduct. In short, Mr. Holmes, you would go far before you found a more dangerous man than Peter Carey, and I have heard that he bore the same character when he commanded his ship. He was known in the trade as Black Peter, and the name was given him, not only on account of his swarthy features and the colour of his huge beard, but

for the humours which were the terror of all around him. I need not say that he was loathed and avoided by every one of his neighbours, and that I have not heard one single word of sorrow about his terrible end.

"You must have read in the account of the inquest about the man's cabin, Mr. Holmes, but perhaps your friend here has not heard of it. He had built himself a wooden outhouse—he always called it the 'cabin'—a few hundred yards from his house, and it was here that he slept every night. It was a little, single-roomed hut, sixteen feet by ten. He kept the key in his pocket, made his own bed, cleaned it himself, and allowed no other foot to cross the threshold. There are small windows on each side, which were covered by curtains and never opened. One of these windows was turned towards the high road, and when the light burned in it at night the folk used to point it out to each other and wonder what Black Peter was doing in there. That's the window, Mr. Holmes, which gave us one of the few bits of positive evidence that came out at the inquest.

"You remember that a stonemason, named Slater, walking from Forest Row about one o'clock in the morning—two days before the murder—stopped as he passed the grounds and looked at the square of light still shining among the trees. He swears that the shadow of a man's head turned sideways was clearly visible on the blind, and that this shadow was certainly not that of Peter Carey, whom he knew well. It was that of a bearded man, but the beard was short and bristled forward in a way very different from that of the captain. So he says, but he had been two hours in the public-house, and it is some distance from the road to the window. Besides, this refers to the Monday, and the crime was done upon the Wednesday.

"On the Tuesday, Peter Carey was in one of his blackest moods, flushed with drink and as savage as a dangerous wild beast. He roamed about the house, and the women ran for it when they heard him coming. Late in the evening, he went down to his own hut. About two o'clock the following morning, his daughter, who slept with her window open, heard a most fearful yell from that direction, but it was no unusual thing for him to bawl and shout when he was in drink, so no notice was taken. On rising at seven, one of the maids noticed that the door of the hut was open, but so great was the terror which the man caused that it was midday before anyone would venture down to see what had become of him. Peeping into the open door, they saw a sight which sent them flying, with white faces, into the village. Within an hour, I was on the spot and had taken over the case.

"Well, I have fairly steady nerves, as you know, Mr. Holmes, but I give you my word, that I got a shake when I put my head into that little house. It was droning like a harmonium with the flies and bluebottles, and the floor and walls were like a slaughter-house. He had called it a cabin, and a cabin it was, sure enough, for you would have thought that you were in a ship. There was a bunk at one end, a sea-chest, maps and charts, a picture of the *Sea Unicorn*, a line of logbooks on a shelf, all exactly as one would expect to find it in a captain's room. And there, in the middle of it was the man him-self—his face twisted like a lost soul in torment, and his great brindled beard stuck upward in his agony. Right through his broad breast a steel harpoon had been driven, and it had sunk deep into the wood of the wall behind him. He was pinned like a beetle on a card. Of course, he was quite dead, and had been so from the instant that he had uttered that last yell of agony.

35   "I know your methods, sir, and I applied them. Before I permitted anything to be moved, I examined most carefully the ground outside, and also the floor of the room. There were no footmarks."

"Meaning that you saw none."

"I assure you, sir, that there were none."

"My good Hopkins. I have investigated many crimes, but I have never yet seen one which was committed by a dying creature. As long as the criminal remains upon two legs so long must there be some indentation, some abrasion, some trifling displacement which can be detected by the scientific searcher. It is incredible that this blood-bespattered room contained no trace which could have aided us. I understand, however, from the inquest that there were some objects which you failed to overlook?"

The young inspector winced at my companion's ironical com-ments.

40   "I was a fool not to call you in at the time, Mr. Holmes. However, that's past praying for now. Yes, there were several objects in the room which called for special attention. One was the harpoon with which the deed was committed. It had been snatched down from a rack on the wall. Two others remained there, and there was a vacant place for the third. On the stock was engraved 'SS. *Sea Unicorn,*

Dundee.' This seemed to establish that the crime had been done in a moment of fury, and that the murderer had seized the first weapon which came in his way. The fact that the crime was committed at two in the morning, and yet Peter Carey was fully dressed, suggested that he had an appointment with the murderer, which is borne out by the fact that a bottle of rum and two dirty glasses stood upon the table."

"Yes," said Holmes; "I think that both inferences are permissible. Was there any other spirit but rum in the room?"

"Yes, there was a tantalus containing brandy and whisky on the sea-chest. It is of no importance to us, however, since the decanters were full, and it had therefore not been used."

"For all that, its presence has some significance," said Holmes. "However, let us hear some more about the objects which do seem to you to bear upon the case."

"There was this tobacco-pouch upon the table."

"What part of the table?"

"It lay in the middle. It was of coarse sealskin—the straight-haired skin, with a leather thong to bind it. Inside was 'P.C.' on the flap. There was half an ounce of strong ship's tobacco in it."

"Excellent! What more?"

Stanley Hopkins drew from his pocket a drab-covered notebook. The outside was rough and worn, the leaves discoloured. On the first page were written the initials "J.H.N." and the date "1853." Holmes laid it on the table and examined it in his minute way, while Hopkins and I gazed over each shoulder. On the second page were the printed letters "C.P.R.," and then came several sheets of numbers. Another heading was "Argentine," another "Costa Rica," and another "San Paulo," each with pages of signs and figures after it.

"What do you make of these?" asked Holmes.

"They appear to be lists of Stock Exchange securities. I thought that 'J.H.N.' were the initials of a broker, and that 'C.P.R.' may have been his client."

"Try Canadian Pacific Railway," said Holmes.

Stanley Hopkins swore between his teeth, and struck his thigh with his clenched hand.

"What a fool I have been!" he cried. "Of course, it is as you say. Then 'J.H.N.' are the only initials we have to solve. I have already examined the old Stock Exchange lists, and I can find no one in

45

50

1883, either in the house or among the outside brokers, whose initials correspond with these. Yet I feel that the clue is the most important one that I hold. You will admit, Mr. Holmes, that there is a possibility that these initials are those of the second person who was present—in other words, of the murderer. I would also urge that the introduction into the case of a document relating to large masses of valuable securities gives us for the first time some indication of a motive for the crime."

Sherlock Holmes's face showed that he was thoroughly taken aback by this new development.

55   "I must admit both your points," said he. "I confess that this notebook, which did not appear at the inquest, modifies any views which I may have formed. I had come to a theory of the crime in which I can find no place for this. Have you endeavoured to trace any of the securities here mentioned?"

"Inquiries are now being made at the offices, but I fear that the complete register of the stockholders of these South American concerns is in South America, and that some weeks must elapse before we can trace the shares."

Holmes had been examining the cover of the notebook with his magnifying lens.

"Surely there is some discolouration here," said he.

"Yes, sir, it is a blood-stain. I told you that I picked the book off the floor."

60   "Was the blood-stain above or below?"

"On the side next the boards."

"Which proves, of course, that the book was dropped after the crime was committed."

"Exactly, Mr. Holmes. I appreciated that point, and I conjectured that it was dropped by the murderer in his hurried flight. It lay near the door."

"I suppose that none of these securities have been found among the property of the dead man?"

65   "No, sir."

"Have you any reason to suspect robbery?"

"No, sir. Nothing seemed to have been touched."

"Dear me, it is certainly a very interesting case. Then there was a knife, was there not?"

"A sheath-knife, still in its sheath. It lay at the feet of the dead man. Mrs. Carey has identified it as being her husband's property."

Holmes was lost in thought for some time. 70

"Well," said he, at last, "I suppose I shall have to come out and have a look at it."

Stanley Hopkins gave a cry of joy.

"Thank you, sir. That will, indeed, be a weight off my mind."

Holmes shook his finger at the inspector.

"It would have been an easier task a week ago," said he. "But 75 even now my visit may not be entirely fruitless. Watson, if you can spare the time, I should be very glad of your company. If you will call a four-wheeler, Hopkins, we shall be ready to start for Forest Row in a quarter of an hour."

Alighting at the small wayside station, we drove for some miles through the remains of widespread woods, which were once part of that great forest which for so long held the Saxon invaders at bay—the impenetrable "weald," for sixty years the bulwark of Britain. Vast sections of it have been cleared, for this is the seat of the first iron-works of the country, and the trees have been felled to smelt the ore. Now the richer fields of the North have absorbed the trade, and nothing save these ravaged groves and great scars in the earth show the work of the past. Here, in a clearing upon the green slope of a hill, stood a long, low, stone house, approached by a curving drive running through the fields. Nearer the road, and surrounded on three sides by bushes, was a small outhouse, one window and the door facing in our direction. It was the scene of the murder.

Stanley Hopkins led us first to the house, where he introduced us to a haggard, gray-haired woman, the widow of the murdered man, whose gaunt and deep-lined face, with the furtive look of terror in the depths of her red-rimmed eyes, told of the years of hardship and ill-usage which she had endured. With her was her daughter, a pale, fair-haired girl, whose eyes blazed defiantly at us as she told us that she was glad that her father was dead, and that she blessed the hand which had struck him down. It was a terrible household that Black Peter Carey had made for himself, and it was with a sense of relief that we found ourselves in the sunlight again and making our way along a path which had been worn across the fields by the feet of the dead man.

The outhouse was the simplest of dwellings, wooden-walled, shingle-roofed, one window beside the door and one on the farther side. Stanley Hopkins drew the key from his pocket and had stooped to the lock, when he paused with a look of attention and surprise upon his face.

"Someone has been tampering with it," he said.

80   There could be no doubt of the fact. The woodwork was cut, and the scratches showed white through the paint, as if they had been that instant done. Holmes had been examining the window.

"Someone has tried to force this also. Whoever it was has failed to make his way in. He must have been a very poor burglar."

"This is a most extraordinary thing," said the inspector, "I could swear that these marks were not here yesterday evening."

"Some curious person from the village, perhaps," I suggested.

"Very unlikely. Few of them would dare to set foot in the grounds, far less try to force their way into the cabin. What do you think of it, Mr. Holmes?"

85   "I think that fortune is very kind to us."

"You mean that the person will come again?"

"It is very probable. He came expecting to find the door open. He tried to get in with the blade of a very small penknife. He could not manage it. What would he do?"

"Come again next night with a more useful tool."

"So I should say. It will be our fault if we are not there to receive him. Meanwhile, let me see the inside of the cabin."

90   The traces of the tragedy had been removed, but the furniture within the little room still stood as it had been on the night of the crime. For two hours, with most intense concentration, Holmes examined every object in turn, but his face showed that his quest was not a successful one. Once only he paused in his patient investigation.

"Have you taken anything off this shelf, Hopkins?"

"No, I have moved nothing."

"Something has been taken. There is less dust in this corner of the shelf than elsewhere. It may have been a book lying on its side. It may have been a box. Well, well, I can do nothing more. Let us walk in these beautiful woods, Watson, and give a few hours to the

birds and the flowers. We shall meet you here later, Hopkins, and
see if we can come to closer quarters with the gentleman who has
paid this visit in the night."

It was past eleven o'clock when we formed our little ambuscade.
Hopkins was for leaving the door of the hut open, but Holmes was
of the opinion that this would rouse the suspicions of the stranger.
The lock was a perfectly simple one, and only a strong blade was
needed to push it back. Holmes also suggested that we should wait,
not inside the hut, but outside it, among the bushes which grew
round the farther window. In this way we should be able to watch
our man if he struck a light, and see what his object was in this
stealthy nocturnal visit.

It was a long and melancholy vigil, and yet brought with it   95
something of the thrill which the hunter feels when he lies beside
the water-pool, and waits for the coming of the thirsty beast of prey.
What savage creature was it which might steal upon us out of the
darkness? Was it a fierce tiger of crime, which could only be taken
fighting hard with flashing fang and claw, or would it prove to be
some skulking jackal, dangerous only to the weak and unguarded?

In absolute silence we crouched amongst the bushes, waiting for
whatever might come. At first the steps of a few belated villagers,
or the sound of voices from the village, lightened our vigil, but one
by one these interruptions died away, and an absolute stillness fell
upon us, save for the chimes of the distant church, which told us of
the progress of the night, and for the rustle and whisper of a fine
rain falling amid the foliage which roofed us in.

Half-past two had chimed, and it was the darkest hour which
precedes the dawn, when we all started as a low but sharp click came
from the direction of the gate. Someone had entered the drive.
Again there was a long silence, and I had begun to fear that it was
a false alarm, when a stealthy step was heard upon the other side of
the hut, and a moment later a metallic scraping and clinking. The
man was trying to force the lock. This time his skill was greater or
his tool was better, for there was a sudden snap and the creak of the
hinges. Then a match was struck, and next instant the steady light
from a candle filled the interior of the hut. Through the gauze
curtain our eyes were all riveted upon the scene within.

The nocturnal visitor was a young man, frail and thin, with a
black moustache, which intensified the deadly pallor of his face. He

could not have been much above twenty years of age. I have never seen any human being who appeared to be in such a pitiable fright, for his teeth were visibly chattering, and he was shaking in every limb. He was dressed like a gentleman, in Norfolk jacket and knickerbockers, with a cloth cap upon his head. We watched him staring round with frightened eyes. Then he laid the candle-end upon the table and disappeared from our view into one of the corners. He returned with a large book, one of the logbooks which formed a line upon the shelves. Leaning on the table, he rapidly turned over the leaves of this volume until he came to the entry which he sought. Then, with an angry gesture of his clenched hand, he closed the book, replaced it in the corner, and put out the light. He had hardly turned to leave the hut when Hopkins's hand was on the fellow's collar, and I heard his loud gasp of terror as he understood that he was taken. The candle was relit, and there was our wretched captive, shivering and cowering in the grasp of the detective. He sank down upon the sea-chest, and looked helplessly from one of us to the other.

"Now, my fine fellow," said Stanley Hopkins, "who are you, and what do you want here?"

100　　The man pulled himself together, and faced us with an effort at self-composure.

"You are detectives, I suppose?" said he. "You imagine I am connected with the death of Captain Peter Carey. I assure you that I am innocent."

"We'll see about that," said Hopkins. "First of all, what is your name?"

"It is John Hopley Neligan."

I saw Holmes and Hopkins exchange a quick glance.

105　　"What are you doing here?"

"Can I speak confidentially?"

"No, certainly not."

"Why should I tell you?"

"If you have no answer, it may go badly with you at the trial."

110　　The young man winced.

"Well, I will tell you," he said. "Why should I not? And yet I hate to think of this old scandal gaining a new lease of life. Did you ever hear of Dawson and Neligan?"

I could see, from Hopkins's face, that he never had, but Holmes was keenly interested.

"You mean the West Country bankers," said he. "They failed for a million, ruined half the county families of Cornwall, and Neligan disappeared."

"Exactly. Neligan was my father."

At last we were getting something positive, and yet it seemed a long gap between an absconding banker and Captain Peter Carey pinned against the wall with one of his own harpoons. We all listened intently to the young man's words.

"It was my father who was really concerned. Dawson had retired. I was only ten years of age at the time, but I was old enough to feel the shame and horror of it all. It has always been said that my father stole all the securities and fled. It is not true. It was his belief that if he were given time in which to realize them, all would be well and every creditor paid in full. He started in his little yacht for Norway just before the warrant was issued for his arrest. I can remember that last night, when he bade farewell to my mother. He left us a list of the securities he was taking, and he swore that he would come back with his honour cleared, and that none who had trusted him would suffer. Well, no word was ever heard from him again. Both the yacht and he vanished utterly. We believed, my mother and I, that he and it, with the securities that he had taken with him, were at the bottom of the sea. We had a faithful friend, however, who is a business man, and it was he who discovered some time ago that some of the securities which my father had with him had reappeared on the London market. You can imagine our amazement. I spent months in trying to trace them, and at last, after many doubtings and difficulties, I discovered that the original seller had been Captain Peter Carey, the owner of this hut.

"Naturally, I made some inquiries about the man. I found that he had been in command of a whaler which was due to return from the Arctic seas at the very time when my father was crossing to Norway. The autumn of that year was a stormy one, and there was a long succession of southerly gales. My father's yacht may well have been blown to the north, and there met by Captain Peter Carey's ship. If that were so, what had become of my father? In any case, if I could prove from Peter Carey's evidence how these securities came on the market it would be a proof that my father had not sold them, and that he had no view to personal profit when he took them.

"I came down to Sussex with the intention of seeing the captain, but it was at this moment that his terrible death occurred. I read at the inquest a description of his cabin, in which it stated that the old logbooks of his vessel were preserved in it. It struck me that if I could see what occurred in the month of August, 1883, on board the *Sea Unicorn*, I might settle the mystery of my father's fate. I tried last night to get at these logbooks, but was unable to open the door. To-night I tried again and succeeded, but I find that the pages which deal with that month have been torn from the book. It was at that moment I found myself a prisoner in your hands."

"Is that all?" asked Hopkins.

120   "Yes, that is all." His eyes shifted as he said it.

"You have nothing else to tell us?"

He hesitated.

"No, there is nothing."

"You have not been here before last night?"

125   "No."

"Then how do you account for *that?*" cried Hopkins, as he held up the damning notebook, with the initials of our prisoner on the first leaf and the blood-stain on the cover.

The wretched man collapsed. He sank his face in his hands, and trembled all over.

"Where did you get it?" he groaned. "I did not know. I thought I had lost it at the hotel."

"That is enough," said Hopkins, sternly. "Whatever else you have to say, you must say in court. You will walk down with me now to the police-station. Well, Mr. Holmes, I am very much obliged to you and to your friend for coming down to help me. As it turns out your presence was unnecessary, and I would have brought the case to this successful issue without you, but, none the less, I am grateful. Rooms have been reserved for you at the Brambletye Hotel, so we can all walk down to the village together."

130   "Well, Watson, what do you think of it?" asked Holmes, as we travelled back next morning.

"I can see that you are not satisfied."

"Oh, yes, my dear Watson, I am perfectly satisfied. At the same time, Stanley Hopkins's methods do not commend themselves to me. I am disappointed in Stanley Hopkins. I had hoped for better things from him. One should always look for a possible alternative, and provide against it. It is the first rule of criminal investigation."

"What, then, is the alternative?"

"The line of investigation which I have myself been pursuing. It may give us nothing. I cannot tell. But at least I shall follow it to the end."

Several letters were waiting for Holmes at Baker Street. He snatched one of them up, opened it, and burst out into a triumphant chuckle of laughter. 135

"Excellent, Watson! The alternative develops. Have you telegraph forms? Just write a couple of messages for me: 'Sumner, Shipping Agent, Ratcliff Highway. Send three men on, to arrive ten to-morrow morning.—Basil.' That's my name in those parts. The other is: 'Inspector Stanley Hopkins, 46 Lord Street, Brixton. Come breakfast to-morrow at nine-thirty. Important. Wire if unable to come.—Sherlock Holmes.' There, Watson, this infernal case has haunted me for ten days. I hereby banish it completely from my presence. To-morrow, I trust that we shall hear the last of it forever."

Sharp at the hour named Inspector Stanley Hopkins appeared, and we sat down together to the excellent breakfast which Mrs. Hudson had prepared. The young detective was in high spirits at his success.

"You really think that your solution must be correct?" asked Holmes.

"I could not imagine a more complete case."

"It did not seem to me conclusive." 140

"You astonish me, Mr. Holmes. What more could one ask for?"

"Does your explanation cover every point?"

"Undoubtedly. I find that young Neligan arrived at the Brambletye Hotel on the very day of the crime. He came on the pretence of playing golf. His room was on the ground-floor, and he could get out when he liked. That very night he went down to Woodman's Lee, saw Peter Carey at the hut, quarrelled with him, and killed him with the harpoon. Then, horrified by what he had done, he fled out of the hut, dropping the notebook which he had brought with him in order to question Peter Carey about these different securities.

You may have observed that some of them were marked with ticks, and the others—the great majority—were not. Those which are ticked have been traced on the London market, but the others, presumably, were still in the possession of Carey, and young Neligan, according to his own account, was anxious to recover them in order to do the right thing by his father's creditors. After his flight he did not dare to approach the hut again for some time, but at last he forced himself to do so in order to obtain the information which he needed. Surely that is all simple and obvious?"

Holmes smiled and shook his head.

145 "It seems to me to have only one drawback, Hopkins, and that is that it is intrinsically impossible. Have you tried to drive a harpoon through a body? No? Tut, tut, my dear sir, you must really pay attention to these details. My friend Watson could tell you that I spent a whole morning in that exercise. It is no easy matter, and requires a strong and practised arm. But this blow was delivered with such violence that the head of the weapon sank deep into the wall. Do you imagine that this anaemic youth was capable of so frightful an assault? Is he the man who hobnobbed in rum and water with Black Peter in the dead of the night? Was it his profile that was seen on the blind two nights before? No, no, Hopkins, it is another and more formidable person for whom we must seek."

The detective's face had grown longer and longer during Holmes's speech. His hopes and his ambitions were all crumbling about him. But he would not abandon his position without a struggle.

"You can't deny that Neligan was present that night, Mr. Holmes. The book will prove that. I fancy that I have evidence enough to satisfy a jury, even if you are able to pick a hole in it. Besides, Mr. Holmes, I have laid my hand upon *my* man. As to this terrible person of yours, where is he?"

"I rather fancy that he is on the stair," said Holmes, serenely. "I think, Watson, that you would do well to put that revolver where you can reach it." He rose and laid a written paper upon a side-table. "Now we are ready," said he.

There had been some talking in gruff voices outside, and now Mrs. Hudson opened the door to say that there were three men inquiring for Captain Basil.

150 "Show them in one by one," said Holmes.

The first who entered was a little Ribston pippin of a man, with ruddy cheeks and fluffy white side-whiskers. Holmes had drawn a letter from his pocket.

"What name?" he asked.

"James Lancaster."

"I am sorry, Lancaster, but the berth is full. Here is half a sovereign for your trouble. Just step into this room and wait there for a few minutes."

The second man was a long, dried-up creature, with lank hair 155
and sallow cheeks. His name was Hugh Pattins. He also received his dismissal, his half-sovereign, and the order to wait.

The third applicant was a man of remarkable appearance. A fierce bull-dog face was framed in a tangle of hair and beard, and two bold, dark eyes gleamed behind the cover of thick, tufted, overhung eyebrows. He saluted and stood sailor-fashion, turning his cap round in his hands.

"Your name?" asked Holmes.

"Patrick Cairns."

"Harpooner?"

"Yes, sir. Twenty-six voyages." 160

"Dundee, I suppose?"

"Yes, sir."

"And ready to start with an exploring ship?"

"Yes, sir."

"What wages?" 165

"Eight pounds a month."

"Could you start at once?"

"As soon as I get my kit."

"Have you your papers?"

"Yes, sir." He took a sheaf of worn and greasy forms from his 170
pocket. Holmes glanced over them and returned them.

"You are just the man I want," said he. "Here's the agreement on the side-table. If you sign it the whole matter will be settled."

The seaman lurched across the room and took up the pen.

"Shall I sign here?" he asked, stooping over the table.

Holmes leaned over his shoulder and passed both hands over his neck.

"This will do," said he. 175

I heard a click of steel and a bellow like an enraged bull. The next instant Holmes and the seaman were rolling on the ground together. He was a man of such gigantic strength that, even with the handcuffs which Holmes had so deftly fastened upon his wrists, he would have very quickly overpowered my friend had Hopkins and I not rushed to his rescue. Only when I pressed the cold muzzle of the revolver to his temple did he at last understand that resistance was vain. We lashed his ankles with cord, and rose breathless from the struggle.

"I must really apologize, Hopkins," said Sherlock Holmes. "I fear that the scrambled eggs are cold. However, you will enjoy the rest of your breakfast all the better, will you not, for the thought that you have brought your case to a triumphant conclusion."

Stanley Hopkins was speechless with amazement.

"I don't know what to say, Mr. Holmes," he blurted out at last, with a very red face. "It seems to me that I have been making a fool of myself from the beginning. I understand now, what I should never have forgotten, that I am the pupil and you are the master. Even now I see what you have done, but I don't know how you did it or what it signifies."

180 "Well, well," said Holmes, good-humouredly. "We all learn by experience, and your lesson this time is that you should never lose sight of the alternative. You were so absorbed in young Neligan that you could not spare a thought to Patrick Cairns, the true murderer of Peter Carey."

The hoarse voice of the seaman broke in on our conversation.

"See here, mister," said he, "I make no complaint of being man-handled in this fashion, but I would have you call things by their right names. You say I murdered Peter Carey, I say I *killed* Peter Carey, and there's all the difference. Maybe you don't believe what I say. Maybe you think I am just slinging you a yarn."

"Not at all," said Holmes. "Let us hear what you have to say."

"It's soon told, and, by the Lord, every word of it is truth. I knew Black Peter, and when he pulled out his knife I whipped a harpoon through him sharp, for I knew that it was him or me. That's how he died. You can call it murder. Anyhow, I'd as soon die with a rope round my neck as with Black Peter's knife in my heart."

185 "How came you there?" asked Holmes.

"I'll tell it you from the beginning. Just sit me up a little, so as I can speak easy. It was in '83 that it happened—August of that year.

Peter Carey was master of the *Sea Unicorn*, and I was spare harpooner. We were coming out of the ice-pack on our way home, with
head winds and a week's southerly gale, when we picked up a little
craft that had been blown north. There was one man on her—a
landsman. The crew had thought she would founder and had made
for the Norwegian coast in the dinghy. I guess they were all
drowned. Well, we took him on board, this man, and he and the
skipper had some long talks in the cabin. All the baggage we took
off with him was one tin box. So far as I know, the man's name was
never mentioned, and on the second night he disappeared as if he
had never been. It was given out that he had either thrown himself
overboard or fallen overboard in the heavy weather that we were
having. Only one man knew what had happened to him, and that
was me, for, with my own eyes, I saw the skipper tip up his heels
and put him over the rail in the middle watch of a dark night, two
days before we sighted the Shetland Lights.

"Well, I kept my knowledge to myself, and waited to see what
would come of it. When we got back to Scotland it was easily hushed
up, and nobody asked any questions. A stranger died by accident,
and it was nobody's business to inquire. Shortly after Peter Carey
gave up the sea, and it was long years before I could find where he
was. I guessed that he had done the deed for the sake of what was
in that tin box, and that he could afford now to pay me well for
keeping my mouth shut.

"I found out where he was through a sailor man that had met
him in London, and down I went to squeeze him. The first night he
was reasonable enough, and was ready to give me what would make
me free of the sea for life. We were to fix it all two nights later.
When I came, I found him three parts drunk and in a vile temper.
We sat down and we drank and we yarned about old times, but the
more he drank the less I liked the look on his face. I spotted that
harpoon upon the wall, and I thought I might need it before I was
through. Then at last he broke out at me, spitting and cursing, with
murder in his eyes and a great clasp-knife in his hand. He had not
time to get it from the sheath before I had the harpoon through
him. Heavens! what a yell he gave! and his face gets between me
and my sleep. I stood there, with his blood splashing round me, and
I waited for a bit, but all was quiet, so I took heart once more. I

looked round, and there was the tin box on the shelf. I had as much right to it as Peter Carey, anyhow, so I took it with me and left the hut. Like a fool I left my baccy-pouch upon the table.

"Now I'll tell you the queerest part of the whole story. I had hardly got outside the hut when I heard someone coming, and I hid among the bushes. A man came slinking along, went into the hut, gave a cry as if he had seen a ghost, and legged it as hard as he could run until he was out of sight. Who he was or what he wanted is more than I can tell. For my part I walked ten miles, got a train at Tunbridge Wells, and so reached London, and no one the wiser.

190   "Well, when I came to examine the box I found there was no money in it, and nothing but papers that I would not dare to sell. I had lost my hold on Black Peter and was stranded in London without a shilling. There was only my trade left. I saw these advertisements about harpooners, and high wages, so I went to the shipping agents, and they sent me here. That's all I know, and I say again that if I killed Black Peter, the law should give me thanks, for I saved them the price of a hempen rope."

"A very clear statement," said Holmes, rising and lighting his pipe. "I think, Hopkins, that you should lose no time in conveying your prisoner to a place of safety. This room is not well adapted for a cell, and Mr. Patrick Cairns occupies too large a proportion of our carpet."

"Mr. Holmes," said Hopkins, "I do not know how to express my gratitude. Even now I do not understand how you attained this result."

"Simply by having the good fortune to get the right clue from the beginning. It is very possible if I had known about this notebook it might have led away my thoughts, as it did yours. But all I heard pointed in the one direction. The amazing strength, the skill in the use of the harpoon, the rum and water, the sealskin tobacco-pouch with the coarse tobacco—all these pointed to a seaman, and one who had been a whaler. I was convinced that the initials 'P.C.' upon the pouch were a coincidence, and not those of Peter Carey, since he seldom smoked, and no pipe was found in his cabin. You remember that I asked whether whisky and brandy were in the cabin. You said they were. How many landsmen are there who would drink rum when they could get these other spirits? Yes, I was certain it was a seaman."

"And how did you find him?"

"My dear sir, the problem had become a very simple one. If it ₁₉₅ were a seaman, it could only be a seaman who had been with him on the *Sea Unicorn.* So far as I could learn he had sailed in no other ship. I spent three days in writing to Dundee, and at the end of that time I had ascertained the names of the crew of the *Sea Unicorn* in 1883. When I found Patrick Cairns among the harpooners, my research was nearing its end. I argued that the man was probably in London, and that he would desire to leave the country for a time. I therefore spent some days in the East End, devised an Arctic expedition, put forth tempting terms for harpooners who would serve under Captain Basil—and behold the result!"

"Wonderful!" cried Hopkins. "Wonderful!"

"You must obtain the release of young Neligan as soon as possible," said Holmes. "I confess that I think you owe him some apology. The tin box must be returned to him, but, of course, the securities which Peter Carey has sold are lost forever. There's the cab, Hopkins, and you can remove your man. If you want me for trial, my address and that of Watson will be somewhere in Norway—I'll send particulars later."

STUDY QUESTIONS

1. The basic structure of Holmes's argument that Patrick Cairns killed Black Peter is as follows:
   (1) The killer was a skilled harpooner
   (2) The killer was a sailor on the *Sea Unicorn* in 1883
   (3) The killer's initials were P.C.
   (4) The only harpooner who sailed on the *Sea Unicorn* in 1883 with initials P.C. was Patrick Cairns
   (5) Patrick Cairns was the killer

$$\frac{(1) + (2) + (3) + (4)}{(5)}$$

What evidence does Holmes have to support (1)–(4)?

2. Find one instance of inductive reasoning in the story, and one example of deductive reasoning.
3. How strong is Holmes's case? If Patrick Cairns had not confessed, do you think a jury would have convicted him?

Reference Chapters: 7, 11, 15

JOHN HENRY WIGMORE

The Borden Case

John Henry Wigmore (1863–1943), professor of law at Northwestern University, was a legal scholar who specialized in the law of evidence. In the following essay, Wigmore presents the facts of the famous Lizzie Borden murder case.

1   On the 4th of August, 1892, was committed in the city of Fall River, Massachusetts, the double murder for which Lizzie Andrew Borden was tried in the month of June, 1893, at New Bedford. Not since the trial of Professor Webster for the murder of Dr. Parkman has such widespread popular interest been aroused; but on this occasion the notoriety far exceeded that of the Webster case, and the report of the proceedings was daily telegraphed to all parts of the country. If we look for the circumstances which made the case such a special theme of discussion, they seem to be three: first, the particularly

[John H. Wigmore, "The Borden Case." *American Law Review* 37 (1893), pp. 806–814.]

brutal mode in which the killing was done; next, the sex of the accused person and her standing in the community; but principally the fact that the evidence was purely circumstantial and was such as to afford singularly conflicting inferences.

In August, 1892, Andrew Jackson Borden was a retired merchant   2
of Fall River, and lived in a house on the east side of Second Street in that city, an important thoroughfare running north and south and faced partly by dwelling houses, partly by business structures. South of the Borden house and closely adjoining was Dr. Kelly's; north of it Mrs. Churchill's; in the rear, but diagonally, Dr. Chagnon's. Mr. Borden was seventy years of age. He was reputed to be worth $300,000 or more, but his family lived in the thrifty and unpretentious style characteristic of New England. The members of the household were Mr. Borden and four others: 1. Mrs. Borden, a short but heavy person, sixty-four years of age, formerly Abby Durfee Gray, now for twenty-five years the second wife of Mr. Borden: 2. Emma Borden, forty-one years of age, a daughter of Mr. Borden's first marriage, and unmarried; 3. Lizzie Andrew Borden, thirty-two years of age, the other child of the first marriage, also unmarried; 4. Bridget Sullivan, a servant who had been with the family nearly three years. Mr. Borden's first wife had died some twenty-eight years before; by the second marriage there was no issue living.

In the latter part of July Emma Borden went to visit friends in   3
Fairhaven, an adjacent town. On Wednesday, August 3, however, the number in the household was restored by a brief visit from John V. Morse, a brother of the first wife. He came just after noon, left for a few hours, returned in the evening, sleeping in the house, and went out the next morning. On Tuesday night, August 2, Mr. and Mrs. Borden were taken suddenly ill with a violent vomiting illness; Lizzie Borden was also slightly affected; Bridget Sullivan was not. On Wednesday morning Mrs. Borden consulted a physician as to this illness. On Thursday morning, August 4, the only persons known to be in the house were Mr. and Mrs. Borden, Miss Borden, Mr. Morse, and the servant Bridget Sullivan. Before describing the occurrences of the morning it is necessary to explain the arrangement of the house.

The appended plan shows the situation of the rooms on the   4 ground and upper floors.[1] As to the ground floor, it is enough to call attention to the fact that there were three doors only: the front door, the kitchen door, and the cellar door; that access from the back door to the front hall might be obtained through the kitchen only, and thence through the sitting-room, or through the dining-room and one or both other rooms, and that in the front hall were two small closets. On the upper floor a doorless partition divided into two small rooms the space over the dining-room. Mr. and Mrs. Borden occupied the room over the kitchen; Lizzie Borden the room over the sitting-room and the front half of the partitioned rooms; and the room over the parlor was used as a guest-room and sewing-room. The door between the rooms of Lizzie Borden and Mr. and Mrs. Borden was permanently locked on both sides (on one by a hook, on the other by a bolt); so that there was no access from the rear part of the upper floor to the front part. Furthermore, the door between the guest-room and Lizzie Borden's room was permanently locked on both sides, and in the latter room a desk stood against the door. In the upper hall over the front door was a clothes closet. As to the condition of the doors below, on August 3 and 4, (1) the front door was locked on Wednesday night by Lizzie Borden, the last one to enter it; the fastening being a spring latch, a bolt, and an ordinary lock; (2) the cellar door (opening into the yard) had been closed on Tuesday and was found locked on Thursday at noon; (3) the kitchen door was locked by Bridget Sullivan on Wednesday night, when she came in (and was found locked by her), but on Thursday morning there was passing in and out, and its condition was not beyond doubt, as we shall see; (4) the door from the bedroom of the Borden couple leading down-stairs was kept locked in their absence from the room. As to the disposition of the inmates of the house on Wednesday, Mr. Morse slept in the guest-chamber, Mr. and Mrs. Borden and Miss Borden in their respective rooms, Bridget Sullivan in the attic at the rear.

5   On Thursday morning shortly after 6, Bridget Sullivan came down the back stairs, got fuel from the cellar, built the fire, and took in the milk. The kitchen door was thus unlocked, the wooden door being left open, the wire screen door fastened, as usual. Just before 7, Mrs. Borden came down. Then Mr. Borden came down, went out

1. See p. 82—Eds.

and emptied his slop-pail, and unlocked the barn door. Mr. Morse then came down, and shortly after 7 the three eat breakfast. Mr. Morse left the house at a quarter before 8, Mr. Borden letting him out and locking the door behind him. Lizzie Borden shortly afterwards came down and began her breakfast in the kitchen. At this point Mr. Borden went upstairs to his room, and Bridget went out in the yard, having an attack of vomiting. After a few minutes' absence she returned and found Lizzie Borden absent, Mrs. Borden dusting the dining-room, and Mr. Borden apparently gone down town. Mrs. Borden then directed Bridget to wash the windows on both sides, and left the kitchen, remarking that she had made the bed in the guestroom and was going up to put two pillow-cases on the pillows there. This was the last time that she was seen alive by any witness. Mr. Borden had left the house somewhere between 9 and 9:30.

Bridget then set to work at the windows, after getting her implements from the cellar, and here the kitchen door seems to have been unlocked and left so. In cleaning the windows of the sitting-room and the dining-room Bridget found nobody present, both Lizzie Borden and Mrs. Borden being elsewhere. As Bridget went out, Lizzie came to the back door, apparently to hook it; but Bridget seems to have dissuaded her. The washing began with the outside of the windows; Bridget proceeded from the two sitting-room windows (where the screen door, now unlocked, was out of sight) to the parlor-front windows, the parlor side window, and the dining-room windows; and during this time neither Lizzie Borden nor Mrs. Borden appeared on the lower floor. Then Bridget entered by the screen door, hooking it behind her, and proceeded to the washing of the inside of the windows, following the same order as before. While washing the first, some one was heard at the front door. Mr. Borden had come home, and failing to enter the screen door, had come round to the front and was trying the door with his key, but the triple fastening prevented his entrance, and Bridget came and opened it before he was obliged to ring the bell. At this moment a laugh or other exclamation was heard from the daughter on the floor above. She came down shortly to the dining-room where Mr. Borden was, asked if there was any mail, and then volunteered the information, "Mrs. Borden has gone out; she had a note from somebody." It was now 10:45, though by a bare possibility 7 or 8 minutes earlier. Mr. Borden took his key, went up the back stairs (the only way to his room), and came down again just as Bridget had finished the second sitting-room window and was passing to the dining-room. Mr. Borden then sat down in the sitting-room; Bridget began on the dining-room windows; and Lizzie Borden put an ironing-board on the dining-room table and began to iron handkerchiefs. This conversation ensued:—

7   "She said, 'Maggie, are you going out this afternoon?' I said, 'I don't know; I might and I might not; I don't feel very well.' She says, 'If you go out, be sure and lock the door, for Mrs. Borden has gone on a sick call, and I might go out too.' Says I, 'Miss Lizzie, who is sick?' 'I don't know; she had a note this morning; it must be in town.'"

8   Then Bridget, finishing the windows, washed out the cloths in the kitchen; and, while she was there, Lizzie Borden stopped her ironing, came into the kitchen and said:—

9   "There is a cheap sale of dress goods at Sargent's to-day at 8 cents a yard."

10   And Bridget said, "I am going to have one."

11   At this point Bridget went upstairs and lay down. In perhaps 3 or 4 minutes the City Hall clock struck, and Bridget's watch showed it to be 11 o'clock. Lizzie Borden never finished her ironing. Miss Russell testified (without contradiction) that she afterwards carried the handkerchiefs upstairs, and that there were 4 or 5 finished with 2 or 3 only sprinkled and ready to iron.

12   The next incident was a cry from below, coming 10 or 15 minutes later:—

13   "Miss Lizzie hollered: 'Maggie, come down.' I said, 'What is the matter?' She says, 'Come down quick, father's dead. Somebody's come in and killed him.'"

14   Bridget hurried down-stairs and found the daughter at the back entrance, leaning against the open wooden door, with her back to the screen door. The daughter sent her for Dr. Bowen, and next, on returning, for her friend Miss Russell, Dr. Bowen being absent. While Miss Russell was being sought, Dr. Bowen and the neighbor, Mrs. Churchill, came, the latter first. Mrs. Churchill gave the alarm at a stable near by, and the telephone message reached police headquarters at 11:15. When Bridget came back and mutual suggestion began, as Bridget relates:—

"I says, 'Lizzie, if I knew where Mrs. Whitehead was I would go   15
and see if Mrs. Borden was there and tell her that Mr. Borden was
very sick.' She says: 'Maggie, I am almost positive I heard her
coming in. Won't you go upstairs to see?' I said: 'I am not going
upstairs alone.'"

Mrs. Churchill offered to go with her. They went upstairs, and   16
as Mrs. Churchill passed up, the door of the guest-room being open,
she saw the clothing of a woman on the floor, the line of sight
running under the bed. She ran on into the room and, standing at
the foot of the bed, saw the dead body of Mrs. Borden stretched on
the floor.[1] It may here be mentioned that the medical testimony
showed, from the temperature of the body, the color and consistency
of the blood, and the condition of the stomach's contents, that Mrs.
Borden's death had occurred between one and two hours earlier,
probably one and one-half hours earlier, than Mr. Borden's,—or not
much later or earlier than 9:30.

During this time the other neighbors were with Lizzie Borden,   17
who had thrown herself on the lounge in the dining-room, not
having been to see her father's or her stepmother's body at any time
since the call for Bridget. At a neighbor's suggestion she went
upstairs to her room, and here without suggestion she afterwards
(within half an hour of the killing) changed her dress and put on a
pink wrapper.

Something must now be said in brief description of the manner   18
in which the two victims had met their death. Mr. Borden's head
bore ten wounds from a cutting instrument wielded with a swing;
the body bore no other injury. The shortest cut was one-half inch
long, the longest was four and one-half inches. Four penetrated the
brain, the skull at the points of penetration being about one-six-
teenth inch thick. The body was found, lying on the right side on
the sofa in the sitting-room, the head nearest the front door, and the
wounds indicated that the assailant stood at or near the head of the
couch and struck down vertically from that direction. Spots of blood
were upon the wall over the sofa (30 to 100), on a picture on the
same wall (40 to 50), on the kitchen door near his feet, and on the
parlor door. On the carpet in front of the sofa, and on a small table
near by, there was no blood. On Mrs. Borden's head and neck (and

1. See plan, p. 82.

not elsewhere) were twenty-two injuries, three ordinary head contusions from falling and nineteen wounds from blows by a cutting instrument,—of these, one was on the back of the neck and eighteen on the head. The shortest was one-half inch, the longest three and one-half inches in length. Four were on the left half of the head, one being a flap wound made in the flesh by a badly-aimed cut from in front. Some thirteen of these made a hole in the top of the skull, crushing into the brain, this part of Mrs. Borden's skull being about one-eighth inch in thickness and the thinnest part of her skull. There were blood spots on the north wall, on the dressing-case (over 75), and on the east wall. The weapon or weapons employed were apparently hatchets or axes. Upon the premises that day were found two hatchets and two axes. Of these only one offered any opportunity for connection with the killings, for the others had handles so marked with ragged portions that they could not have been cleansed from the blood which they must have received. Of the fourth some mention will be made later.

19   On Tuesday, Wednesday and Thursday, August 9, 10 and 11, the inquest was held by Judge Blaisdell, and on Thursday evening Lizzie Borden was arrested on charge of committing the murders. The preliminary trial began before Judge Blaisdell, August 25, continuing until September 1, when she was found probably guilty and ordered to be held for the grand jury. The indictment was duly found, and on Monday, June 5, 1893, the trial began in the Superior Court of Bristol County, at the New Bedford Court House. In accord with the law of the State, the Court for such a trial was composed of three judges of the Superior Court of the Commonwealth. Those who officiated on this occasion were Mason, C. J., Blodgett, J., and Dewey, J.

20   The case for the prosecution was conducted by Hosea M. Knowlton, District Attorney for the County,[2] and Wm. H. Moody, District Attorney of Essex County.[3] The case for the defence was conducted by George D. Robinson,[4] Melvin O. Adams,[5] and Andrew J. Jennings.[6]

2. Afterwards Justice of the Massachusetts Supreme Court.
3. Afterwards Justice of the United States Supreme Court.
4. Former Governor of Massachusetts.
5. Eminent at the Boston Bar in the defense of criminal cases.
6. Former partner of Mr. Justice Morton of the Massachusetts Supreme Court.

We now come to consider the question, what points did the   21
prosecution attempt to make against Lizzie Borden in charging the
crime upon her? It endeavored to show, *first*, prior indications, *(a)*
Motive, *(b)* Design; *second*, concomitant indications, *(a)* Opportunity, *(b)* Means and Capacity; *third*, posterior indications, *(a)* Consciousness of Guilt. Let us take these in order very briefly.

1. *(a) Motive*. The family history was brought in to show that the   22
accused was not on the best of terms with her stepmother. This was
evidenced by the testimony of: (1) A dressmaker, who reported that
in a conversation held some time previously, when her "mother"
was mentioned, she answered: "Don't say 'mother' to me. She is a
mean, good-for-nothing old thing. We do not have much to do with
her; I stay in my room most of the time." "Why, you come down to
your meals?" "Yes, sometimes; but we don't eat with them if we can
help it." (2) The servant, who reported that, though she never saw
any quarreling, "most of the time they did not eat with the father
and mother." (3) The uncle, who did not see Lizzie Borden during
the visit from Wednesday noon till Thursday noon: (4) the sister,
Emma, who explained the ill-feeling partly on the ground of a small
transfer of property by the father to his wife a few years before, and
reported that since that time the accused had ceased saying
"mother" and addressed her as "Mrs. Borden," and that a gift of
other property to the daughters had only partially allayed the
ill-feeling; (5) the police officer, who on asking Lizzie Borden on
Thursday noon, "When did you last see your mother?" was answered, "She is not my mother. My mother is dead." The general
effect of the motive testimony purported to be that the daughters
were afraid of the property going to the second wife, to their
exclusion, and that this fomented an ill-feeling existing on more or
less general grounds of incompatibility.

(b) Design. No evidence was offered of a specific design to kill   23
with the weapons used. But it was attempted to show a general
intention to get rid of the victims: (1) Testimony of a druggist and
of by-standers as to an attempted purchase of prussic acid in the
forenoon of Wednesday, the day before the killing:—

"This party came in there and inquired if I kept prussic acid. I   24
was standing out there; I walked in ahead. She asked me if we kept
prussic acid. I informed her that we did. She asked me if she could
buy ten cents' worth of me. I informed her that we did not sell

prussic acid unless by a physician's prescription. She then said that she had bought this several times, I think; I think she said several times before. I says: 'Well, my good lady, it is something we don't sell unless by a prescription from the doctor, as it is a very dangerous thing to handle.' I understood her to say she wanted it to put on the edge of a seal-skin cape, if I remember rightly. She did not buy anything, no drug at all, no medicine? No, sir." This was excluded, for reasons to be mentioned later.

25   (2) Testimony of a conversation on the same Wednesday, during an evening call on Miss Russell, an intimate friend:—

26   The prisoner said: "I have made up my mind, Alice, to take your advice and go to Marion, and I have written there to them that I shall go, but I cannot help feeling depressed; I cannot help feeling that something is going to happen to me; I cannot shake it off. Last night," she said, "we were all sick; Mr. and Mrs. Borden were quite sick and vomited; I did not vomit, and we are afraid that we have been poisoned; the girl did not eat the baker's bread and we did, and we think it may have been the baker's bread."

27   "No," said Miss Russell, "if it had been that, some other people would have been sick in the same way."

28   "Well, it might have been the milk; our milk is left outside upon the steps."

29   "What time is your milk left?"

30   "At 4 o'clock in the morning."

31   "It is light then, and no one would dare to come in and touch it at that time."

32   "Well," said the prisoner, "probably that is so. But father has been having so much trouble with those with whom he has dealings that I am afraid some of them will do something to him; I expect nothing but that the building will be burned down over our heads. The barn has been broken into twice."

33   "That," said Miss Russell, "was merely boys after pigeons."

34   "Well, the house has been broken into in broad daylight when Maggie and Emma and I were the only ones in the house. I saw a man the other night when I went home lurking about the buildings, and as I came he jumped and ran away. Father had trouble with a man the other day about a store. There were angry words, and he turned him out of the house."

(3) The suggestion to Bridget that she should go to town and purchase the dress-goods mentioned.

2. *(a) Opportunity.* One of the chief efforts of the prosecution was to prove an exclusive opportunity on the part of the accused. The essential result of the testimony bearing on this may be gleaned from what has already been noted. *(b) Means* and *Capacity.* The medical testimony showed that there was nothing in the assaults which a woman of her strength might not have accomplished. The lengthy testimony in regard to the fourth hatchet was directed to showing that it was not incapable of being the weapon used. The handle was broken off; but the presence of ashes on the handle in all other places but the broken end, as well as the appearance of the break, showed that it was a fresh one, and not impossibly one made after the killing; and if thus made, it was not impossible that the hatchet was used in killing, washed, rubbed in ashes, broken off, and the fragment burnt. A strong effort was made by the defense to discredit these results, which rested chiefly on the reports of police officers, but it had little effect.

3. *(a) Consciousness of Guilt.* This, with exclusive opportunity, were the main objects of the prosecution's attack. Much that was here offered was excluded, and this exclusion possibly affected the result of the case. The points attempted to be shown were: (1) Falsehoods to prevent detection of the first death; (2) falsehoods as to the doings of the accused; (3) knowledge of the first death; (4) concealment of knowledge of the first death; (5) destruction of suspicious materials.

(1) To Bridget and to her father the accused said, as already related, that her mother had received a note and gone out. The same statement she made to Mrs. Churchill and to Marshal Fleet. No note, however, was found; no one who brought a note or sent a note came forward or was heard of; no sound or sight of the sort was perceived by Bridget or any others. The only blot upon an almost perfectly conducted trial was the attempt of the counsel for the defense in argument to show that the information as to the note emanated originally from Bridget and that the accused merely repeated it. This was decidedly a breach of propriety, because it was not merely an argument suggesting the fair possibility of that explanation, but a distinct assertion that the testimony was of that purport, and,

therefore, in effect, a false quotation of the testimony. In truth the accused's statement about the note was her own alone and was one of the facts to be explained.

39   (2) Here were charged three falsehoods: *(a)* When the accused was asked where she was at the time of the killing of Mr. Borden, she said that she went out to the barn (to Dr. Bowen) "looking for some iron or irons," (to Miss Russell) "for a piece of iron or tin to fix a screen," (to the mayor and an officer and at the coroner's inquest)[7] in the barn loft, eating some pears and "looking over lead for sinkers." The inconsistency of the explanations was offered as very suggestive. The day was shown to be a very hot one, and the loft was argued to be too hot for such a sojourn. Moreover Officer Medley testified to going into the barn, in the loft, and finding the floor covered with dust, easily taking an impression from his hand or foot, but on his arrival quite devoid of any traces of the previous presence of another. The trustworthiness of his statements was attacked by witnesses who said that they and others had been there before the officer. The priority of their visits was not placed beyond doubt; but the effect of the officer's statement of course fell from practical proof to a merely probative circumstance.

40   *(b)* When the accused was describing her discovery of the father's death, she said (to Officer Mullaly) that she heard "a peculiar noise, something like a scraping noise, and came in and found the door open;" (to the servant) that she heard a groan and rushed in and found her father; (to Mrs. Churchill) that she heard a distress noise, came in, and found her father; (at the inquest) that after eating pears in the loft and looking over lead, she came down, returned to the kitchen, looked in the stove to see if the fire was hot enough for her ironing, found that it was not, put her hat down, started to go upstairs and wait for Bridget's noon-day fire, and thus discovered her father; (to Officer Harrington) that she was up in the loft of the barn and thus did not hear any outcry or noise of any kind; (to Marshal Hilliard) that after half an hour up in the barn, she came in and found her father. Here, again, a substantial inconsistency was charged.

41   *(c)* Mr. Borden had on, when found, a pair of congress boots or

7. Her inquest testimony was excluded, for reasons to be considered later.

gaiters; but at the inquest the accused, before this was pointed out, testified that when he came home about 10:45, she assisted him to lie down on the sofa, took off his boots, and put on his slippers.

(3) Her knowledge of the first death was said to have been indicated: *(a)* By the inevitable discovery of the body in the guest-room through the open door, or of the murderer either in passing about or in going up and down the stairs; *(b)* by the noise of the scuffle, if another had done it, and by the thud of the heavy woman's fall; *(c)* by the readiness with which the accused suggested that Mrs. Borden must have returned;[8] (1) for as her father had been in the room off the hall from 10:45 to, say, 11, and as she had been out in the barn from 11 till the killing was discovered and others came in, there was no time when the mother could have returned since the father's return, and up to that time the accused herself predicated her absence.

(4) If this knowledge existed, then beyond doubt the concealment of it and the pretense of ignorance involved in sending Bridget to get the step-mother was strongly indicative of guilt.

(5) Some attempt was made to show a degree of secrecy and obstruction to official investigation of the rooms; but with little or no result. On Sunday morning, however (the officers having informed her on Saturday that she was suspected of the crime), when Emma Borden and Lizzie Borden were in the kitchen and officers were in the yard, Alice Russell came in:—

"I saw Miss Lizzie at the other end of the stove, I saw Miss Emma at the sink. Miss Lizzie was at the stove and she had a skirt in her hand, and her sister turned and said: 'What are you going to do?' and Lizzie said, 'I am going to burn this old thing up; it is covered with paint.' I left the room then, and on coming back, Miss Lizzie stood up toward the cupboard door, and she appeared to be either ripping something down or tearing part of this garment. I said to her: 'I wouldn't let anybody see me do that, Lizzie.' She didn't make any answer, but just stepped one step farther back, up toward the cupboard door. . . . Afterwards, I said to them, 'I am afraid, Lizzie, the worst thing you could have done was to burn that dress. I have

8. This, however, was not argued at the trial. Moreover, no attempt was made to show that Mrs. Borden had no latch-key to the knowledge of the accused.

been asked about your dress.' She said: 'Oh, what made you let me do it? Why didn't you tell me?'"

46   The prosecution naturally attempted, first, to identify this dress as the one worn on the morning of the killing; in this they failed; second, to show at least that the dress worn on that day was missing, and was not the one handed over by the accused, as the dress of that morning. On this point they made out a very strong case. The dress handed over by the accused to the officers as the one worn on Thursday morning, while ironing, and afterwards, was a silk dress, of a dark blue effect; the testimony, however, pointed strongly to the wearing of a cotton dress, light blue with a dark figure. Such a dress existed, and had been worn on the day before, but not on Friday or Saturday.

47   Thus far the prosecution. The defense began with character evidence based on the accused's cooperation in Sunday-school and charitable work and her good standing as a church member. The motive-evidence was not shaken; though the sister of the accused represented the ill-feeling to be of minimum intensity. The design-evidence of prussic acid did not come to the jury. In regard to exclusive opportunity, the defense made no break in the chain of the prosecution, except in showing that the screen door was not closed at all moments during the morning. The evidence as to the possibility of an unseen escape from the house was not potent on either side. But no traces of another person were shown within the house; and no suspicious person was located in the vicinity of the house—if we except some vague reports of a tramp, of a pale, excited young man, and the like, being seen on the street, near by, within a day or an hour of the killing. The attempt failed to show the impossibility of the handleless hatchet having been used—unless we assume (what the defense desired to suggest) that the testimony of all the officers was wilfully false. Coming to the evidence of consciousness of guilt,—the defense could not shake the story of the note; they merely suggested that it might have been a part of the scheme of the murderer to divert suspicion. They searched for the note and they advertised for the sender or carrier, but nothing appeared. The inconsistent stories about going to the barn were explained by the excitement of the moment; the inquest-story—with the most marked divergence—was excluded. Lead was

found in the loft; but no fish-line was shown[9] and no screen was identified. It was suggested that perhaps both explanations were true, that both purposes co-existed. The inconsistent stories as to her return and discovery of the murder were in part slid over, in part ignored, and in part discredited.[10]

The discrepancy between the statement about the slippers and the actual foot-coverings did not get to the jury. As to the circumstances indicating knowledge, their force was a matter of argument and probability merely; the defense urged the contrary hypotheses which suggest themselves to all. The dress burning was explained by the sister to have taken place in consequence of a suggestion of hers; but Miss Russell's testimony contradicted this. The defense offered to show a custom in the family of burning all old dresses, but this was rejected. Another offer, also rejected, was to show the conduct of a demented-looking man, seen in the woods near the town, a few days after the murder, carrying an axe, and exclaiming "Poor Mrs. Borden!" 48

The stronghold of the defense was the utter absence of all such traces or marks as would presumably be found upon the murderer. No blood was seen upon her by the five or six persons who came in within ten minutes and before she donned the pink wrapper. No garment was found with blood or other traces upon it.[11] No weapon bearing blood or other traces was found within or without the house. One or two of the experts were willing to say that it was practically impossible to deal the twenty-nine blows without receiving more or less blood on the garments and perhaps in the hair (though it does not appear that her head was examined for blood). It is safe to say that this was the decisive fact of the case. 49

It is, of course, impossible to rehearse here all the minor details of evidence and argument offered on either side. It has been necessary to make a summary estimate of the force of certain evidence mentioned. 50

9. The lead-for-sinkers statement had not been admitted, but the counsel for the defense took it up in his argument.

10. The inquest-story, going into particulars, had never been admitted; but there were still at least two distinct statements.

11. Except a white skirt having at the back and below a spot of blood as large as a pinhead, the spot being otherwise explainable.

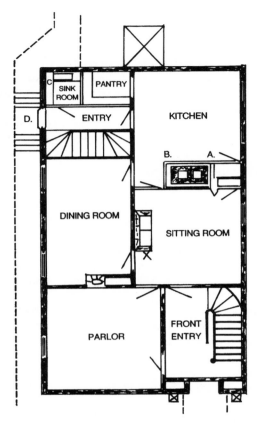

—Adapted from Porter's Fall River Tragedy

GROUND FLOOR, BORDEN HOUSE

Mr. Borden's head rested on sofa arm by door—see X—as he took his nap.

A. Lizzie stood here, in angle of coal-closet door, as she burned the dress she had placed on the closet shelf some hours before.

B. Stove in which she burned it.

C. Where Emma was washing the dishes when Lizzie began to burn the dress.

Note that upstairs guest room (5, next floor plan) can be reached from kitchen entry (D) only by way of the kitchen and sitting room. Shelves at left side of sink room and two sides of pantry; sitting-room closet also filled with shelves. Extremely small broom closet facing stairs at right side of front door is not indicated on this floor plan.

On Tuesday, June 20, at 4:32 in the afternoon, after less than an
hour and a half of deliberation, the jury returned a verdict of "not
guilty."[12]

STUDY QUESTIONS

1. Wigmore classifies the evidence in the trial into several
   categories. Do these categories provide an exhaustive classification for the evidence relevant to a murder charge? Which
   category of evidence, in general, seems strongest?
2. For each category of evidence, determine whether (a) Lizzie
   Borden's guilt is consistent with all the evidence, and (b)
   whether it is the only hypothesis consistent with all the
   evidence.
3. Diagram the prosecution's overall argument, showing how
   the various categories of evidence integrate into a case against
   Lizzie Borden.
4. Do you think Lizzie Borden killed her parents?
5. If you had been on her jury, would you have voted to convict
   her?

Reference Chapters: 2, 7, 11

12. It was reported that they were of one mind on the first ballot, and remained
an hour in general conversation, at the suggestion of one member, merely to avoid
letting the counsel for the Commonwealth suppose that his argument did not
receive consideration.

SIGMUND FREUD

Human Nature Is Inherently Bad

Sigmund Freud (1856–1939) is famous for being the founder of psychoanalysis and for his view of sex as the primary motivating factor in human behavior. He studied medicine at the University of Vienna, and after graduation researched the organic bases of mental disorders. Freud's research led him to the study of hypnosis, which in turn led him to investigate the possibility that our mental lives are shaped primarily by unconscious forces.

1   [M]en are not gentle creatures who want to be loved, and who at the most can defend themselves if they are attacked; they are, on the contrary, creatures among whose instinctual endowments is to be reckoned a powerful share of aggressiveness. As a result, their neighbour is for them not only a potential helper or sexual object, but also someone who tempts them to satisfy their aggressiveness on him, to exploit his capacity for work without compensation, to use him sexually without his consent, to seize his possessions, to humiliate him, to cause him pain, to torture and to kill him. *Homo homini lupus.*[1] Who, in the face of all his experience of life and of history, will have the courage to dispute this assertion? As a rule this cruel aggressiveness waits for some provocation or puts itself at the service of some other purpose, whose goal might also have been

[Sigmund Freud, excerpt from *Civilization and Its Discontents*, transl. James Strachey. New York: W. W. Norton, 1962, pp. 58–59.]

1. Man is a wolf to man.—Eds.

reached by milder measures. In circumstances that are favourable to it, when the mental counter-forces which ordinarily inhibit it are out of action, it also manifests itself spontaneously and reveals man as a savage beast to whom consideration towards his own kind is something alien. Anyone who calls to mind the atrocities committed during the racial migrations or the invasions of the Huns, or by the people known as Mongols under Jenghiz Khan and Tamerlane, or at the capture of Jerusalem by the pious Crusaders, or even, indeed, the horrors of the recent World War—anyone who calls these things to mind will have to bow humbly before the truth of this view.

The existence of this inclination to aggression, which we can   2
detect in ourselves and justly assume to be present in others, is the factor which disturbs our relations with our neighbour and which forces civilization into such a high expenditure [of energy]. In consequence of this primary mutual hostility of human beings, civilized society is perpetually threatened with disintegration. The interest of work in common would not hold it together; instinctual passions are stronger than reasonable interests. Civilization has to use its utmost efforts in order to set limits to man's aggressive instincts and to hold the manifestations of them in check by psychical reaction-formations. Hence, therefore, the use of methods intended to incite people into identifications and aim-inhibited relationships of love, hence the restriction upon sexual life, and hence too the ideal's commandment to love one's neighbour as oneself—a commandment which is really justified by the fact that nothing else runs so strongly counter to the original nature of man. In spite of every effort, these endeavours of civilization have not so far achieved very much. It hopes to prevent the crudest excesses of brutal violence by itself assuming the right to use violence against criminals, but the law is not able to lay hold of the more cautious and refined manifestations of human aggressiveness. . . .

STUDY QUESTIONS

   1. Does Freud argue inductively or deductively for his conclusion that humans have an innate "inclination to aggression"?

What role does his mentioning of the invasions of the Huns and World War I play in his argument?

2. What does Freud consider "the original nature of man"?

3. Freud states that "instinctual passions are stronger than reasonable interests." Does the existence of long-standing cooperative institutions weaken Freud's claim? Would Freud have to concede that humans also have a "cooperative instinct"?

4. Can you think of explanations for historical atrocities that do not depend upon the existence of human instincts? Does Freud say anything in this passage to rule out such possibilities?

Reference Chapters: 15, 18

PART THREE

On Poetry, Proverbs, and Childish Insults

JOHN ENRIGHT

What Is Poetry?

John Enright, born in 1952, is a computer consultant living in Chicago. Many attempts have been made to define poetry, and the difficulty of the task can be seen in the fact that none of the attempts has met with widespread acceptance. In the following selection, Enright critiques a few of those previous definitions and defends his own.

Poetry, among the arts, has a history of being poorly, even mysteriously, defined. Part of the problem is that many of those offering definitions have been poets; and too many of their definitions have

1

[John Enright, excerpt from "What Is Poetry," *Objectively Speaking* (Autumn 1989).]

been more poetical than precise. Emily Dickinson, for instance, on being asked her criterion for poetry, wrote: "[i]f I read a book and it makes my whole body so cold that no fire can warm me, I know that it is poetry. If I feel physically as if the top of my head were taken off, I know that is poetry." This is vivid and forceful, but it tells us much more about Emily Dickinson than it does about poetry.

2   Dylan Thomas called poetry " . . . the rhythmic, inevitably narrative movement from an over-clothed blindness to a naked vision." In his inclusion of the word "rhythmic," Thomas's definition is a step up from Dickinson's, for he indicates one of poetry's distinguishing marks.

3   An all-too-common failing of proposed definitions of poetry is that they could apply equally well to other art forms. Witness Shelley's: "[p]oetry is the record of the best and happiest moments of the best and happiest minds." Poe did better: "I would define the poetry of words as the rhythmical creation of beauty." This excludes most of the other arts, but does not sharply distinguish poetry from *song*, which also uses words and rhythm.

4   A formal definition combines a genus and a differentia—the general class to which a thing belongs, and the characteristics that make it different from the rest of the things in that class.

5   The proper genus of poetry is art form. We differentiate art forms from one another by the specific material media of the forms. The medium of poetry is language, but novels and vocal songs also depend upon language. The unique medium of poetry is *language utilizing the musical elements intrinsic to the language.* In contrast, prose makes little use of language's musical potential, and song turns upon a musical element which is extrinsic to language: melody.

6   Two classical definitions of poetry, "musical speech" and "rhythmical speech," are not far off the mark. The trouble with "musical speech" is that it does not differentiate poetry from song. The trouble with "rhythmical speech" is that rhythm is *not* the only musical element that poetry employs. There is *much* more to the music of language than beat.

7   An objection to be expected here is that I am simply defining poetry as *verse,* and that I must consequently accept as poetry commercial jingles, such as: "Hold the pickles, hold the lettuce!/Special orders don't upset us." However, the purpose of de-

fining poetry's genus as "art form" was precisely to forestall such classification. An art form must project a deeply held view of life—which the above Burger King jingle does not.

It *is* true that much of modern "poetry" cannot qualify as real poetry by this definition. But I consider this to be a virtue rather than a fault. . . .

8

STUDY QUESTIONS

1. Enright quotes Emily Dickinson, Dylan Thomas, and Percy Bysshe Shelley. Using the criteria for good definitions, critique the quotations as definitions of poetry.
2. Consider several poems that you know. Do all of them utilize *"the musical elements intrinsic to the language,"* as Enright claims?
3. What is the distinction Enright makes between poetry and verse? Do you agree with Enright that the commercial jingle he quotes isn't poetry, merely verse?
4. In what way does Enright's proposed definition of poetry depend on an assumed definition of art?
5. Paragraph 5 states that poetry's genus is art form, and paragraph 7 that an art form must project "a deeply held view of life." This means that Enright's definition of poetry excludes a large body of writing that other people might classify under this concept. How should one decide which attributes are essential to poetry?

Reference Chapters: 2, 3

PETER FARB

Children's Insults

Peter Farb (1929–1980) was a consultant to the Smithsonian Institution, a visiting lecturer in English at Yale University, and a curator of the Riverside Museum in New York.

1   The insults spoken by adults are usually more subtle than the simple name-calling used by children, but children's insults make obvious some of the verbal strategies people carry into adult life. Most parents engage in wishful thinking when they regard name-calling as good-natured fun which their children will soon grow out of. Name-calling is not good-natured and children do not grow out of it; as adults they merely become more expert in its use. Nor is it true that "sticks and stones may break my bones, but names will never hurt me." Names can hurt very much because children seek out the victim's true weakness, then jab exactly where the skin is thinnest. Name-calling can have major impact on a child's feelings about his identity, and it can sometimes be devastating to his psychological development.

2   Almost all examples of name-calling by children fall into four categories:

> 1. Names based on physical peculiarities, such as deformities, use of eyeglasses, racial characteristics, and so forth. A child may be called *Flattop* because he was born with a misshapen skull—or, for obvious reasons, *Fat Lips, Gimpy, Four Eyes, Peanuts, Fatso, Kinky,* and so on.

[Peter Farb, excerpt from *Word Play: What Happens When People Talk.* New York: Alfred A. Knopf, 1973, pp. 74–75.]

2. Names based on a pun or parody of the child's own name. Children with last names like Fitts, McClure, and Farb usually find them converted to *Shits*, *Manure*, and *Fart*.
3. Names based on social relationships. Examples are *Baby* used by a sibling rival or *Chicken Shit* for someone whose courage is questioned by his social group.
4. Names based on mental traits—such as *Clunkhead*, *Dummy*, *Jerk*, and *Smartass*.

These four categories were listed in order of decreasing offensiveness to the victims. Children regard names based on physical peculiarities as the most cutting, whereas names based on mental traits are, surprisingly, not usually regarded as very offensive. Most children are very vulnerable to names that play upon the child's rightful name—no doubt because one's name is a precious possession, the mark of a unique identity and one's masculinity or femininity. Those American Indian tribes that had the custom of never revealing true names undoubtedly avoided considerable psychological damage.

STUDY QUESTIONS

1. What principle of classification does Farb use to establish his categories?
2. Farb offers a classification of children's insults, and also claims that children's insults can have harmful long-term effects. How does he argue for this?
3. Farb says that insulting names "seek out the victim's true weakness, then jab exactly where the skin is thinnest." Do all of Farb's examples do this?
4. Can any of Farb's examples be placed in more than one of the categories he offers?
5. Make a list of other children's insults. Do all of them fit into Farb's categories?

Reference Chapter: 2

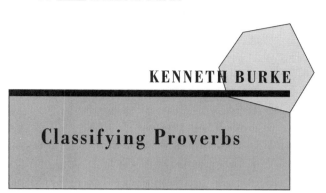

KENNETH BURKE

Classifying Proverbs

Kenneth Burke, born in Pittsburgh in 1897, was an American music and literary critic, sociologist, and writer of fiction and poetry. He took no formal degrees, but did receive eight honorary doctorates and was visiting professor of sociology at Harvard University.

1   Examine random specimens in *The Oxford Dictionary of English Proverbs.* You will note, I think, that there is no "pure" literature here. Everything is "medicine." Proverbs are designed for consolation or vengeance, for admonition or exhortation, for foretelling.

2   Or they name typical, recurrent situations. That is, people find a certain social relationship recurring so frequently that they must "have a word for it." The Eskimos have special names for many different kinds of snow (fifteen, if I remember rightly) because variations in the quality of snow greatly affect their living. Hence, they must "size up" snow much more accurately than we do. And the same is true of social phenomena. Social structures give rise to "type" situations, subtle subdivisions of the relationships involved in competitive and cooperative acts. Many proverbs seek to chart, in more or less homey and picturesque ways, these "type" situations. I submit that such naming is done, not for the sheer glory of the thing, but because of its bearing upon human welfare. A different

[Kenneth Burke, excerpt from "Literature as Equipment for Living," in *The Philosophy of Literary Form: Studies in Symbolic Action*, 2nd ed. Baton Rouge, La.: Louisiana State University Press, 1967, pp. 293–297.]

name for snow implies a different kind of hunt. Some names for snow imply that one should not hunt at all. And similarly, the names for typical, recurrent social situations are not developed out of "disinterested curiosity," but because the names imply a command (what to expect, what to look out for).

To illustrate with a few representative examples: 3

Proverbs designed for consolation: "The sun does not shine on 4 both sides of the hedge at once." "Think of ease, but work on." "Little troubles the eye, but far less the soul." "The worst luck now, the better another time." "The wind in one's face makes one wise." "He that hath lands hath quarrels." "He knows how to carry the dead cock home." "He is not poor that hath little, but he that desireth much."

For vengeance: "At length the fox is brought to the furrier." 5 "Shod in the cradle, barefoot in the stubble." "Sue a beggar and get a louse." "The higher the ape goes, the more he shows his tail." "The moon does not heed the barking of dogs." "He measures another's corn by his own bushel." "He shuns the man who knows him well." "Fools tie knots and wise men loose them."

Proverbs that have to do with foretelling (the most obvious are 6 those to do with the weather): "Sow peas and beans in the wane of the moon, Who soweth them sooner, he soweth too soon." "When the wind's in the north, the skillful fisher goes not forth." "When the sloe tree is as white as a sheet, sow your barley whether it be dry or wet." "When the sun sets bright and clear, An easterly wind you need not fear. When the sun sets in a bank, A westerly wind we shall not want."

In short: "Keep your weather eye open": be realistic about sizing 7 up today's weather, because your accuracy has bearing upon tomor-row's weather. And forecast not only the meteorological weather, but also the social weather: "When the moon's in the full, then wit's in the wane." "Straws show which way the wind blows." "When the fish is caught, the net is laid aside." "Remove an old tree, and it will wither to death." "The wolf may lose his teeth, but never his nature." "He that bites on every weed must needs light on poison." "Whether the pitcher strikes the stone, or the stone the pitcher, it is bad for the pitcher." "Eagles catch no flies." "The more laws, the more offenders."

In this foretelling category we might also include the recipes for 8 wise living, sometimes moral, sometimes technical: "First thrive, and then wive." "Think with the wise but talk with the vulgar." "When the fox preacheth, then beware your geese." "Venture a small fish to catch a great one." "Respect a man, he will do the more."

9   In the class of "typical, recurrent situations" we might put such proverbs and proverbial expressions as: "Sweet appears sour when we pay." "The treason is loved but the traitor is hated." "The wine in the bottle does not quench thirst." "The sun is never the worse for shining on a dunghill." "The lion kicked by an ass." "The lion's share." "To catch one napping." "To smell a rat." "To cool one's heels."

10   By all means, I do not wish to suggest that this is the only way in which the proverbs could be classified. For instance, I have listed in the "foretelling" group the proverb, "When the fox preacheth, then beware your geese." But it could obviously be "taken over" for vindictive purposes. Or consider a proverb like "Virtue flies from the heart of a mercenary man." A poor man might obviously use it either to console himself for being poor (the implication being, "Because I am poor in money I am rich in virtue") or to strike at another (the implication being, "When he got money, what else could you expect of him but deterioration?"). In fact, we could even say that such symbolic vengeance would itself be an aspect of solace. And a proverb like "The sun is never the worse for shining on a dunghill" (which I have listed under "typical recurrent situations") might as well be put in the vindictive category.

11   The point of issue is not to find categories that "place" the proverbs once and for all. What I want is categories that suggest their active nature. Here is no "realism for its own sake." Here is realism for promise, admonition, solace, vengeance, foretelling, instruction, charting, all for the direct bearing that such acts have upon matters of welfare. . . .

12   Proverbs are *strategies* for dealing with *situations.* In so far as situations are typical and recurrent in a given social structure, people develop names for them and strategies for handling them. Another name for strategies might be *attitudes.*

13   People have often commented on the fact that there are *contrary* proverbs. But I believe that the above approach to proverbs suggests

a necessary modification of that comment. The apparent contradictions depend upon differences in *attitude,* involving a correspondingly different choice of *strategy.* Consider, for instance, the *apparently* opposite pair: "Repentance comes too late" and "Never too late to mend." The first is admonitory. It says in effect: "You'd better look out, or you'll get yourself too far into this business." The second is consolatory, saying in effect: "Buck up, old man, you can still pull out of this."

STUDY QUESTIONS

1. What principle does Burke follow in classifying the proverbs he lists?
2. Think of some other proverbs, or look up a representative sample in a reference book. Do they all fit into Burke's classificatory scheme?
3. At the beginning of paragraph 10, Burke says: ". . . I do not wish to suggest that this is the only way in which the proverbs could be classified." What instead does he see as the purpose of classifying the proverbs? Does this have any implications for the principle he is employing?

Reference Chapter: 2

PART FOUR

On Euthanasia and Abortion

JUDGE GAVIN K. LETTS

Satz v. *Perlmutter* [Euthanasia Is Justifiable]

*Gavin K. Letts is a judge in the Fourth District of the Florida
District Court of Appeal. He was educated at Edinburgh University in Scotland and at Washington and Lee Law School.
He practiced law in Florida before coming to the judicial
bench in 1977. The following case deals with voluntary, passive euthanasia. "Voluntary," in this context, means that the
patient knowingly consents to the procedure. "Passive euthanasia" means stopping medical treatment with the intent of allowing a patient to die naturally. It is usually contrasted with
active euthanasia, which involves taking a positive action*

[Judge Gavin K. Letts, *Satz* v. *Perlmutter*, Florida District Court of Appeal,
Fourth District. Reprinted from *Southern Reporter*, 362 So. 2d 160; affirmed 379
So. 2d 359; legal references omitted.]

> *(such as giving the patient an injection) with the intent of caus-*
> *ing the patient's death. Judge Letts concludes that individuals*
> *have the right to refuse medical treatment, even if they know*
> *their deaths will follow.*

Satz v. *Perlmutter*
DECIDED SEPTEMBER 13, 1978
LETTS, JUDGE.

1   The State here appeals a trial court order permitting the removal of an artificial life sustaining device from a competent, but terminally ill adult. . . .

2   Seventy-three year old Abe Perlmutter lies mortally sick in a hospital, suffering from amyotrophic lateral sclerosis (Lou Gehrig's disease) diagnosed in January 1977. There is no cure and normal life expectancy, from time of diagnosis, is but two years. In Mr. Perlmutter, the affliction has progressed to the point of virtual incapability of movement, inability to breathe without a mechanical respirator and his very speech is an extreme effort. Even with the respirator, the prognosis is death within a short time. Notwithstanding, he remains in command of his mental faculties and legally competent. He seeks, with full approval of his adult family, to have the respirator removed from his trachea, which act, according to his physician, based upon medical probability, would result in "a reasonable life expectancy of less than one hour." Mr. Perlmutter is fully aware of the inevitable result of such removal, yet has attempted to remove it for himself (hospital personnel, activated by an alarm, reconnected it). He has repeatedly stated to his family, "I'm miserable, take it out" and at a bedside hearing, told the obviously concerned trial judge that whatever would be in store for him if the respirator were removed, "it can't be worse than what I'm going through now."

3   Pursuant to all of the foregoing, and upon the petition of Mr. Perlmutter himself, the trial judge entered a detailed and thoughtful final judgment which included the following language:

> ORDERED AND ADJUDGED that Abe Perlmutter, in the exercise of his right of privacy, may remain in defendant hospital or leave said hospital, free of the mechanical respirator now attached to his body and all defendants and their staffs are restrained from interfering with Plaintiff's decision.

We agree with the trial judge.

The State's position is that it (1) has an overriding duty to   4
preserve life, and (2) that termination of supportive care, whether
it be by the patient, his family or medical personnel, is an unlawful
killing of a human being under the Florida Murder Statute Section
782.04, Florida Statutes (1977) or Manslaughter under Section
782.08. The hospital, and its doctors, while not insensitive to this
tragedy, fear not only criminal prosecution if they aid in removal
of the mechanical device, but also civil liability. In the absence of
prior Florida law on the subject, their fears cannot be discounted.

The pros and cons involved in such tragedies which bedevil   5
contemporary society, mainly because of incredible advancement
in scientific medicine, are all exhaustively discussed in Superinten-
dent of Belchertown v. Saikewicz, Mass. As *Saikewicz* points out, the
right of an individual to refuse medical treatment is tempered by
the State's:

1. Interest in the preservation of life.
2. Need to protect innocent third parties.
3. Duty to prevent suicide.
4. Requirement that it help maintain the ethical integrity of medical practice.

In the case at bar, none of these four considerations surmount the   6
individual wishes of Abe Perlmutter. Thus we adopt the view of the
line of cases discussed in *Saikewicz* which would allow Abe Perlmutter the right to refuse or discontinue treatment based upon "the
constitutional right to privacy . . . an expression of the sanctity of
individual free choice and self-determination." We would stress
that this adoption is limited to the specific facts now before us,
involving a competent adult patient. The problem is less easy of
solution when the patient is incapable of understanding and we,
therefore, postpone a crossing of that more complex bridge until
such time as we are required to do so.

Preservation of Life

There can be no doubt that the State *does* have an interest in   7
preserving life, but we again agree with *Saikewicz* that "there is a

substantial distinction in the State's insistence that human life be saved where the affliction is curable, as opposed to the State interest where, as here, the issue is not whether, but when, for how long and at what cost to the individual [his] life may be briefly extended." In the case at bar the condition is terminal, the patient's situation wretched and the continuation of his life temporary and totally artificial.

8   Accordingly, we see no compelling State interest to interfere with Mr. Perlmutter's expressed wishes.

Protection of Third Parties

9   Classically, this protection is exemplified in the case Application of the President and Directors of Georgetown College, Inc., where the patient, by refusing treatment, is said to be abandoning his minor child, which abandonment the State as *parens patriae* ["father of the country"] sought to prevent. We point out that Abe Perlmutter is 73, his family adult and all in agreement with his wishes. The facts do not support abandonment.

10   As to suicide, the facts here unarguably reveal that Mr. Perlmutter would die, but for the respirator. The disconnecting of it, far from causing his unnatural death by means of a "death producing agent" in fact will merely result in his death, if at all, from natural causes. The testimony of Mr. Perlmutter, like the victim in the *Georgetown College* case, supra, is that he really wants to live, but to do so, God and Mother Nature willing, under his own power. This basic wish to live, plus the fact that he did not self-induce his horrible affliction, precludes his further refusal of treatment being classed as attempted suicide.

11   Moreover we find no requirement in the law that a competent, but otherwise mortally sick, patient undergo the surgery or treatment which constitutes the only hope for temporary prolongation of his life. This being so, we see little difference between a cancer ridden patient who declines surgery, or chemotherapy, necessary for his temporary survival and the hopeless predicament which tragically afflicts Abe Perlmutter. It is true that the latter appears more drastic because affirmatively, a mechanical device must be discon-

nected, as distinct from mere inaction. Notwithstanding, the principle is the same, for in both instances the hapless, but mentally competent, victim is choosing not to avail himself of one of the expensive marvels of modern medical science.

The State argues that a patient has *no right* to refuse treatment and cites several of the familiar blood transfusion cases. However, a reading of these reveal substantial distinctions between them and the case at bar. In the blood transfusion cases, the patient is either incompetent to make a medical decision, equivocal about making it ("he would not agree to be transfused but would not resist a court order permitting it because it would be the court's will and not his own.") or it is a family member making the decision for an inert or minor third party patient. By contrast, we find, and agree with, several cases upholding the right of a competent adult patient to refuse treatment for himself. From this agreement, we reach our conclusion that, because Abe Perlmutter has a right to refuse treatment in the first instance, he has a concomitant right to discontinue it.

Ethics of Medical Practice

Lastly, as to the ethical integrity of medical practice, we again adopt the language of *Saikewicz:*

> Prevailing medical ethical practice does not, without exception, demand that all efforts toward life prolongation be made in all circumstances. Rather, as indicated in *Quinlan,* the prevailing ethical practice seems to be to recognize that the dying are more often in need of comfort than treatment. Recognition of the right to refuse necessary treatment in appropriate circumstances is consistent with existing medical mores; such a doctrine does not threaten either the integrity of the medical profession, the proper role of hospitals in caring for such patients or the State's interest in protecting the same. It is not necessary to deny a right of self-determination to a patient in order to recognize the interests of doctors, hospitals, and medical personnel in attendance on the patient. . . .

It is our conclusion, therefore, under the facts before us, that when these several public policy interests are weighed against the rights of Mr. Perlmutter, the latter must and should prevail. Abe Perlmutter should be allowed to make his choice to die with dignity, notwithstanding over a dozen legislative failures in this state to adopt suitable legislation in this field. It is all very convenient to insist on continuing Mr. Perlmutter's life so that there can be no question of foul play, no resulting civil liability and no possible trespass on medical ethics. However, it is quite another matter to do so at the patient's sole expense and against his competent will, thus inflicting never ending physical torture on his body until the inevitable, but artificially suspended, moment of death. Such a course of conduct invades the patient's constitutional right of privacy, removes his freedom of choice and invades his right to self-determine.

15   The judgment of the trial court is hereby affirmed. . . .

STUDY QUESTIONS

1. In paragraphs 4 and 5, Letts presents a number of considerations that weigh against terminating Perlmutter's medical treatment. How many distinct considerations are presented there? For example, is (1) in paragraph 4 the same as point 1 in paragraph 5? And is point 3 in paragraph 5 an application of point 1 or a distinct point?

2. In paragraphs 7 through 13, Letts argues against the considerations presented in paragraphs 4 and 5. Reconstruct in standard form each of Letts's arguments.

3. In paragraph 10, Letts argues that Perlmutter's desire to stop medical treatment is not the same as a desire to commit suicide. What definition of "suicide" does Letts's position imply?

4. Letts holds that the act of removing Perlmutter's mechanical respirator should not be classified as murder, manslaughter, or suicide. So how should we classify it? Is it a species of the genus "intentional termination of human life"?

5. Letts holds that the overriding consideration is the individual's right to refuse medical treatment. Does he argue for this premise?

6. In 1973, the American Medical Association (AMA) made the following statement: "The cessation of the employment of extraordinary means to prolong the life of the body when there is irrefutable evidence that biological death is imminent is the decision of the patient and/or his immediate family." Does Perlmutter's case meet the conditions set in the AMA statement?

Reference Chapters: 2, 3, 5, 7

STEPHEN G. POTTS

Looking for the Exit Door: Killing and Caring in Modern Medicine

The term "euthanasia" derives from the Greek eu *(meaning "well") and* thanatos *(meaning "death"). In the following selection, the British physician Stephen G. Potts argues that in spite of the good intentions that may motivate euthanasia, institutionalizing its practice will lead to abuses and dehumanized values.*

[I am opposed] to any attempt to institutionalise euthanasia . . . because the risks of such institutionalisation are so grave as to outweigh the very real suffering of those who might benefit from it.

[Stephen G. Potts, "Looking for the Exit Door: Killing and Caring in Modern Medicine." *Houston Law Review* 25 (1988), pp. 504–509, 510–511.]

Risks of Institutionalisation

2 Among the potential effects of a legalised practice of euthanasia are the following:

3 **1 Reduced Pressure to Improve Curative or Symptomatic Treatment** If euthanasia had been legal forty years ago, it is quite possible that there would be no hospice movement today. The improvement in terminal care is a direct result of attempts made to minimise suffering. If that suffering had been extinguished by extinguishing the patients who bore it, then we may never have known the advances in the control of pain, nausea, breathlessness and other terminal symptoms that the last twenty years have seen.

4 Some diseases that were terminal a few decades ago are now routinely cured by newly developed treatments. Earlier acceptance of euthanasia might well have undercut the urgency of the research efforts which led to the discovery of those treatments. If we accept euthanasia now, we may well delay by decades the discovery of effective treatments for those diseases that are now terminal.

5 **2 Abandonment of Hope** Every doctor can tell stories of patients expected to die within days who surprise everyone with their extraordinary recoveries. Every doctor has experienced the wonderful embarrassment of being proven wrong in their pessimistic prognosis. To make euthanasia a legitimate option as soon as the prognosis is pessimistic enough is to reduce the probability of such extraordinary recoveries from low to zero.

7 **3 Increased Fear of Hospitals and Doctors** Despite all the efforts at health education, it seems there will always be a transference of the patient's fear of illness from the illness to the doctors and hospitals who treat it. This fear is still very real and leads to large numbers of late presentations of illnesses that might have been cured if only the patients had sought help earlier. To institutionalise euthanasia, however carefully, would undoubtedly magnify all the latent fear of doctors and hospitals harbored by the public. The inevitable result would be a rise in late presentations and, therefore, preventable deaths.

8 **4 Difficulties of Oversight and Regulation** Both the Dutch and the Californian proposals [to allow euthanasia in some cases] list sets of precautions designed to prevent abuses. They acknowledge that such abuses are a possibility. I am far from convinced that

the precautions are sufficient to prevent either those abuses that have been foreseen or those that may arise after passage of the law. The history of legal "loopholes" is not a cheering one: Abuses might arise when the patient is wealthy and an inheritance is at stake, when the doctor has made mistakes in diagnosis and treatment and hopes to avoid detection, when insurance coverage for treatment costs is about to expire, and in a host of other circumstances.

5 Pressure on the Patient Both sets of proposals seek to limit the influence of the patient's family on the decision, again acknowledging the risks posed by such influence. Families have all kinds of subtle ways, conscious and unconscious, of putting pressure on a patient to request euthanasia and relieve them of the financial and social burden of care. Many patients already feel guilty for imposing burdens on those who care for them, even when the families are happy to bear that burden. To provide an avenue for the discharge of that guilt in a request for euthanasia is to risk putting to death a great many patients who do not wish to die.

6 Conflict with Aims of Medicine The pro-euthanasia movement cheerfully hands the dirty work of the actual killing to the doctors who, by and large, neither seek nor welcome the responsibility. There is little examination of the psychological stresses imposed on those whose training and professional outlook are geared to the saving of lives by asking them to start taking lives on a regular basis. Euthanasia advocates seem very confident that doctors can be relied on to make the enormous efforts sometimes necessary to save some lives, while at the same time assenting to requests to take other lives. Such confidence reflects, perhaps, a high opinion of doctors' psychic robustness, but it is a confidence seriously undermined by the shocking rates of depression, suicide, alcoholism, drug addiction, and marital discord consistently recorded among this group.

7 Dangers of Societal Acceptance It must never be forgotten that doctors, nurses, and hospital administrators have personal lives, homes, and families, or that they are something more than just doctors, nurses or hospital administrators. They are *citizens* and a significant part of the society around them. I am very worried about what the institutionalisation of euthanasia will do to society, in general, and, particularly how much it will further erode our attachment to the sixth commandment. ["Thou shalt not kill."]

How will we regard murderers? What will we say to the terrorist who justifies killing as a means to his political end when we ourselves justify killing as a means to a humanitarian end? I do not know and I daresay the euthanasia advocates do not either, but I worry about it and they appear not to. They need to justify their complacency.

12   **8 The Slippery Slope**   How long after acceptance of voluntary euthanasia will we hear the calls for nonvoluntary euthanasia? There are thousands of comatose or demented patients sustained by little more than good nursing care. They are an enormous financial and social burden. How soon will the advocates of euthanasia be arguing that we should "assist them in dying"—for, after all, they won't mind, will they?

13   How soon after *that* will we hear the calls for involuntary euthanasia, the disposal of the burdensome, the unproductive, the polluters of the gene pool? We must never forget the way the Nazi euthanasia programme made this progression in a few short years. "Oh, but they were barbarians," you say, and so they were, but not at the outset.

14   If developments in terminal care can be represented by a progression from the CURE mode of medical care to the CARE mode, enacting voluntary euthanasia legislation would permit a further progression to the KILL mode. The slippery slope argument represents the fear that, if this step is taken, then it will be difficult to avoid a further progression to the CULL mode, as illustrated:

> CURE   The central aim of medicine
> CARE   The central aim of terminal care once patients are beyond cure
> KILL   The aim of the proponents of euthanasia for those patients beyond cure and not helped by care
> CULL   The feared result of weakening the prohibition on euthanasia

15   I do not know how easy these moves will be to resist once voluntary euthanasia is accepted, but I have seen little evidence that the modern euthanasia advocates care about resisting them or even worry that they might be possible.

16   **9 Costs and Benefits**   Perhaps the most disturbing risk of all is posed by the growing concern over medical costs. Euthanasia is,

after all, a very cheap service. The cost of a dose of barbiturates and curare and the few hours in a hospital bed that it takes them to act is minute compared to the massive bills incurred by many patients in the last weeks and months of their lives. Already in Britain, there is a serious under-provision of expensive therapies like renal dialysis and intensive care, with the result that many otherwise preventable deaths occur. Legalising euthanasia would save substantial financial resources which could be diverted to more "useful" treatments. These economic concerns already exert pressure to accept euthanasia, and, if accepted, they will inevitably tend to enlarge the category of patients for whom euthanasia is permitted.

Each of these objections could, and should, be expanded and   17
pressed harder. I do not propose to do so now, for it is sufficient for my purposes to list them as *risks,* not inevitabilities. Several elements go into our judgment of the severity of a risk: the *probability* that the harm in question will arise (the odds), the *severity* of the harm in question (the stakes), and the ease with which the harm in question can be corrected (the *reversibility*). The institutionalisation of euthanasia is such a radical departure from anything that has gone before in Western society that we simply cannot judge the probability of any or all of the listed consequences. Nor can we rule any of them out. There must, however, be agreement that the severity of each of the harms listed is enough to horrify. Furthermore, many of the potential harms seem likely to prove very difficult, if not impossible, to reverse by reinstituting a ban on euthanasia.

Weighing the Risks

For all these reasons, the burden of proof *must* lie with those who   18
would have us gamble by legalising euthanasia. They should demonstrate beyond reasonable doubt that the dangers listed will not arise, just as chemical companies proposing to introduce a new drug are required to demonstrate that it is safe as well as beneficial. Thus far, the proponents of euthanasia have relied exclusively on the compassion they arouse with tales of torment mercifully cut short by death, and have made little or no attempt to shoulder the burden

of proving that legalising euthanasia is safe. Until they make such an attempt and carry it off successfully, their proposed legislation must be rejected outright.

The Right to Die and the Duty to Kill

19   The nature of my arguments should have made it clear by now that I object, not so much to individual acts of euthanasia, but to institutionalising it as a practice. All the pro-euthanasia arguments turn on the individual case of the patient in pain, suffering at the center of an intolerable existence. They exert powerful calls on our compassion, and appeal to our pity, therefore, we assent too readily when it is claimed that such patients have a *"right to die"* as an escape from torment. So long as the right to die means no more than the right to refuse life-prolonging treatment and the right to rational suicide, I agree. The advocates of euthanasia want to go much further than this though. They want to extend the right to die to encompass the right to receive assistance in suicide and, beyond that, the right to be killed. Here, the focus shifts from the patient to the agent, and from the killed to the killer; but, the argument begins to break down because our compassion does not extend this far.

20   If it is true that there is a right to be assisted in suicide or a right to be killed, then it follows that someone, somewhere, has a *duty* to provide the assistance or to do the killing. When we look at the proposed legislation, it is very clear upon whom the advocates of euthanasia would place this duty: the doctor. It would be the doctor's job to provide the pills and the doctor's job to give the lethal injection. The regulation of euthanasia is meant to prevent anyone, other than the doctor, from doing it. Such regulation would ensure that the doctor does it with the proper precautions and consultations, and would give the doctor security from legal sanctions for doing it. The emotive appeal of euthanasia is undeniably powerful, but it lasts only so long as we can avoid thinking about who has to do the killing, and where, and when, and how. Proposals to institutionalise euthanasia force us to think hard about these things, and the chill that their contemplation generates is deep enough to freeze any proponent's ardor.

STUDY QUESTIONS

1. Given what you know, rank Potts's nine arguments in order of strength.
2. If we can add up the risks involved in institutionalizing euthanasia, can they be measured against the suffering of individual patients who might be candidates for euthanasia?
3. In veterinary medicine, euthanasia is a common practice for pets that are suffering with no prospect of recovery. Do you think this has led to a decline in the care of pets by veterinarians? Does this provide any analogical evidence for or against Pott's argument?
4. In paragraph 20, Potts says that if there is a *right* to be assisted in suicide, then someone has a *duty* to provide the assistance. What definition of "right" does Potts's claim assume?

Reference Chapters: 3, 7

RONALD REAGAN

Abortion and the Conscience of the Nation

Ronald Reagan, born in 1911 in Tampico, Illinois, was a radio sports announcer, an actor in such films as Knute Rockne, All-

[Ronald Reagan, excerpt from *Abortion and the Conscience of the Nation.* Nashville, Tenn.: Thomas Nelson Publishers, 1984, pp. 15–16, 18–19, 21–25, 27–36, 38.]

American, and The Hasty Heart, *governor of California, and, from 1981 to 1989, the President of the United States. In the following essay, Reagan offers his grounds for opposing the practice of abortion.*

1   The tenth anniversary of the Supreme Court decision in *Roe* v. *Wade* is a good time for us to pause and reflect. Our nationwide policy of abortion-on-demand through all nine months of pregnancy was neither voted for by our people nor enacted by our legislators—not a single state had such unrestricted abortion before the Supreme Court decreed it to be national policy in 1973. But the consequences of this judicial decision are now obvious: since 1973, more than 15 million unborn children have had their lives snuffed out by legalized abortions. That is over ten times the number of Americans lost in all our nation's wars.

2   Make no mistake, abortion-on-demand is not a right granted by the Constitution. No serious scholar, including one disposed to agree with the Court's result, has argued that the framers of the Constitution intended to create such a right. Shortly after the *Roe* v. *Wade* decision, Professor John Hart Ely, now Dean of Stanford Law School, wrote that the opinion "is not constitutional law and gives almost no sense of an obligation to try to be." Nowhere do the plain words of the Constitution even hint at a "right" so sweeping as to permit abortion up to the time the child is ready to be born. Yet that is what the Court ruled.

3   As an act of "raw judicial power" (to use Justice White's biting phrase), the decision by the seven-man majority in *Roe* v. *Wade* has so far been made to stick. But the Court's decision has by no means settled the debate. Instead, *Roe* v. *Wade* has become a continuing prod to the conscience of the nation.

4   Abortion concerns not just the unborn child, it concerns every one of us. The English poet, John Donne, wrote: " . . . any man's death diminishes me, because I am involved in mankind; and therefore never send to know for whom the bell tolls; it tolls for thee."

5   We cannot diminish the value of one category of human life— the unborn—without diminishing the value of all human life. We saw tragic proof of this truism last year[1] when the Indiana courts

___

1. I.e., 1983.—Eds.

allowed the starvation death of "Baby Doe" in Bloomington because the child had Down's Syndrome.

Many of our fellow citizens grieve over the loss of life that has 6 followed *Roe* v. *Wade*. Margaret Heckler, soon after being nominated to head the largest department of our government, Health and Human Services, told an audience that she believed abortion to be the greatest moral crisis facing our country today. And the revered Mother Teresa, who works in the streets of Calcutta ministering to dying people in her world-famous mission of mercy, has said that "the greatest misery of our time is the generalized abortion of children."

Over the first two years of my administration I have closely 7 followed and assisted efforts in Congress to reverse the tide of abortion—efforts of congressmen, senators and citizens responding to an urgent moral crisis. Regrettably, I have also seen the massive efforts of those who, under the banner of "freedom of choice," have so far blocked every effort to reverse nationwide abortion-on-demand.

Despite the formidable obstacles before us, we must not lose 8 heart. This is not the first time our country has been divided by a Supreme Court decision that denied the value of certain human lives. The *Dred Scott* decision of 1857 was not overturned in a day, or a year, or even a decade. At first, only a minority of Americans recognized and deplored the moral crisis brought about by denying the full humanity of our black brothers and sisters; but that minority persisted in their vision and finally prevailed. They did it by appealing to the hearts and minds of their countrymen, to the truth of human dignity under God. From their example, we know that respect for the sacred value of human life is too deeply engrained in the hearts of our people to remain forever suppressed. But the great majority of the American people have not yet made their voices heard, and we cannot expect them to—any more than the public voice arose against slavery—*until* the issue is clearly framed and presented.

What, then, is the real issue? I have often said that when we talk 9 about abortion, we are talking about two lives—the life of the mother and the life of the unborn child. Why else do we call a pregnant woman a mother? I have also said that anyone who doesn't feel sure whether we are talking about a second human life should clearly give life the benefit of the doubt. If you don't know whether a body is alive or dead, you would never bury it. I think this consideration itself should be enough for all of us to insist on protecting the unborn.

10   The case against abortion does not rest here, however, for medical practice confirms at every step the correctness of these moral sensibilities. Modern medicine treats the unborn child as a patient. Medical pioneers have made great breakthroughs in treating the unborn—for genetic problems, vitamin deficiencies, irregular heart rhythms, and other medical conditions. Who can forget George Will's moving account of the little boy who underwent brain surgery six times during the nine weeks before he was born? Who is the *patient* if not that tiny unborn human being who can feel pain when he or she is approached by doctors who come to kill rather than to cure?

11   The real question today is not when human life begins, but, *What is the value of human life?* The abortionist who reassembles the arms and legs of a tiny baby to make sure all its parts have been torn from its mother's body can hardly doubt whether it is a human being. The real question for him and for all of us is whether that tiny human life has a God-given right to be protected by the law—the same right we have.

12   What more dramatic confirmation could we have of the real issue than the Baby Doe case in Bloomington, Indiana? The death of that tiny infant tore at the hearts of all Americans because the child was undeniably a live human being—one lying helpless before the eyes of the doctors and the eyes of the nation. The real issue for the courts was *not* whether Baby Doe was a human being. The real issue was whether to protect the life of a human being who had Down's Syndrome, who would probably be mentally handicapped, but who needed a routine surgical procedure to unblock his esophagus and allow him to eat. A doctor testified to the presiding judge that, even with his physical problem corrected, Baby Doe would have a "nonexistent" possibility for "a minimally adequate quality of life"—in other words, that retardation was the equivalent of a crime deserving the death penalty. The judge let Baby Doe starve and die, and the Indiana Supreme Court sanctioned his decision.

13   Federal law does not allow federally-assisted hospitals to decide that Down's Syndrome infants are not worth treating, much less to

decide to starve them to death. Accordingly, I have directed the Departments of Justice and Health and Human Services to apply civil rights regulations to protect handicapped newborns. All hospitals receiving federal funds must post notices which will clearly state that failure to feed handicapped babies is prohibited by federal law. The basic issue is whether to value and protect the lives of the handicapped, whether to recognize the sanctity of human life. This is the same basic issue that underlies the question of abortion.

The 1981 Senate hearings on the beginning of human life 14 brought out the basic issue more clearly than ever before. The many medical and scientific witnesses who testified disagreed on many things, but not on the *scientific* evidence that the unborn child is alive, is a distinct individual, or is a member of the human species. They did disagree over the *value* question, whether to give value to a human life at its early and most vulnerable stages of existence.

Regrettably, we live at a time when some persons do *not* value 15 all human life. They want to pick and choose which individuals have value. Some have said that only those individuals with "consciousness of self" are human beings. One such writer has followed this deadly logic and concluded that "shocking as it may seem, a newly born infant is not a human being."

A Nobel Prize winning scientist has suggested that if a handi- 16 capped child "were not declared fully human until three days after birth, then all parents could be allowed the choice." In other words, "quality control" to see if newly born human beings are up to snuff.

Obviously, some influential people want to deny that every 17 human life has intrinsic, sacred worth. They insist that a member of the human race must have certain qualities before they accord him or her status as a "human being."

Events have borne out the editorial in a California medical 18 journal which explained three years before *Roe* v. *Wade* that the social acceptance of abortion is a "defiance of the long-held Western ethic of intrinsic and equal value for every human life regardless of its stage, condition, or status."

Every legislator, every doctor, and every citizen needs to recog- 19 nize that the real issue is whether to affirm and protect the sanctity of all human life, or to embrace a social ethic where some human lives are valued and others are not. As a nation, we must choose between the sanctity of life ethic and the "quality of life" ethic.

20   I have no trouble identifying the answer our nation has always given to this basic question, and the answer that I hope and pray it will give in the future. America was founded by men and women who shared a vision of the value of each and every individual. They stated this vision clearly from the very start in the Declaration of Independence, using words that every schoolboy and schoolgirl can recite:

> We hold these truths to be self-evident, that all men are created equal, that they are endowed by their Creator with certain unalienable rights, that among these are life, liberty, and the pursuit of happiness.

21   We fought a terrible war to guarantee that one category of mankind—black people in America—could not be denied the inalienable rights with which their Creator endowed them. The great champion of the sanctity of all human life in that day, Abraham Lincoln, gave us his assessment of the Declaration's purpose. Speaking of the framers of that noble document, he said:

> This was their majestic interpretation of the economy of the Universe. This was their lofty, and wise, and noble understanding of the justice of the Creator to His creatures. Yes, gentlemen, to all His creatures, to the whole great family of man. In their enlightened belief, nothing stamped with the divine image and likeness was sent into the world to be trodden on. . . . They grasped not only the whole race of man then living, but they reached forward and seized upon the farthest posterity. They erected a beacon to guide their children and their children's children, and the countless myriads who should inhabit the earth in other ages.

He warned also of the danger we would face if we closed our eyes to the value of life in any category of human beings:

> I should like to know if taking this old Declaration of Independence, which declares that all men are equal upon principle and making exceptions to it where will it stop. If one man says it does not mean a Negro, why not another say it does not mean some other man?

22   When Congressman John A. Bingham of Ohio drafted the Fourteenth Amendment to guarantee the rights of life, liberty, and property to all human beings, he explained that *all* are "entitled to the protection of American law, because its divine spirit of equality

declares that all men are created equal." He said the rights guaranteed by the amendment would therefore apply to "any human being." Justice William Brennan, in another case decided only the year before *Roe* v. *Wade,* referred to our society as one that "strongly affirms the sanctity of life."

Another William Brennan—not the Justice—has reminded us   23
of the terrible consequences that can follow when a nation rejects the sanctity of life ethic:

> The cultural environment for a human holocaust is present whenever any society can be misled into defining individuals as less than human and therefore devoid of value and respect.

As a nation today, we have *not* rejected the sanctity of human life.   24
The American people have not had an opportunity to express their view on the sanctity of human life in the unborn. I am convinced that Americans do not want to play God with the value of human life. It is not for us to decide who is worthy to live and who is not. Even the Supreme Court's opinion in *Roe* v. *Wade* did not explicitly reject the traditional American idea of intrinsic worth and value in all human life; it simply dodged this issue.

The Congress has before it several measures that would enable   25
our people to reaffirm the sanctity of human life, even the smallest and the youngest and the most defenseless. The Human Life Bill expressly recognizes the unborn as human beings and accordingly protects them as persons under our Constitution. This bill, first introduced by Senator Jesse Helms, provided the vehicle for the Senate hearings in 1981 which contributed so much to our understanding of the real issue of abortion.

The Respect Human Life Act, just introduced in the ninety-   26
eighth Congress, states in its first section that the policy of the United States is "to protect innocent life, both before and after birth." This bill, sponsored by Congressman Henry Hyde and Senator Roger Jepsen, prohibits the federal government from performing abortions or assisting those who do so, except to save the life of the mother. It also addresses the pressing issue of infanticide which, as we have seen, flows inevitably from permissive abortion as another step in the denial of the inviolability of innocent human life.

I have endorsed each of these measures, as well as the more   27 difficult route of constitutional amendment, and I will give these initiatives my full support. Each of them, in different ways, attempts to reverse the tragic policy of abortion-on-demand imposed by the Supreme Court ten years ago. Each of them is a decisive way to affirm the sanctity of human life.

28   We must all educate ourselves to the reality of the horrors taking place. Doctors today know that unborn children can feel a touch within the womb and that they respond to pain. But how many Americans are aware that abortion techniques are allowed today, in all fifty states, that burn the skin of a baby with a salt solution, in an agonizing death that can last for hours?

29   Another example: two years ago, the *Philadelphia Inquirer* ran a Sunday special supplement on "The Dreaded Complication." The "dreaded complication" referred to in the article—the complication feared by doctors who perform abortions—is the *survival* of the child despite all the painful attacks during the abortion procedure. Some unborn children *do* survive the late-term abortions the Supreme Court has made legal. Is there any question that these victims of abortion deserve our attention and protection? Is there any question that those who *don't* survive were living human beings before they were killed?

30   Late-term abortions, especially when the baby survives, but is then killed by starvation, neglect, or suffocation, show once again the link between abortion and infanticide. The time to stop both is now. As my administration acts to stop infanticide, we will be fully aware of the real issue that underlies the death of babies before and soon after birth.

31   Our society has, fortunately, become sensitive to the rights and special needs of the handicapped, but I am shocked that physical or mental handicaps of newborns are still used to justify their extinction. This administration has a Surgeon General, Dr. C. Everett Koop, who has done perhaps more than any other American for handicapped children, by pioneering surgical techniques to help them, by speaking out on the value of their lives, and by working with them in the context of loving families. You will not find his former patients advocating the so-called "quality-of-life" ethic.

32   I know that when the true issue of infanticide is placed before the American people, with all the facts openly aired, we will have no trouble deciding that a mentally or physically handicapped baby

has the same intrinsic worth and right to life as the rest of us. As the New Jersey Supreme Court said two decades ago, in a decision upholding the sanctity of human life, "a child need not be perfect to have a worthwhile life."

Whether we are talking about pain suffered by unborn children, or about late-term abortions, or about infanticide, we inevitably focus on the humanity of the unborn child. Each of these issues is a potential rallying point for the sanctity of life ethic. Once we as a nation rally around any one of these issues to affirm the sanctity of life, we will see the importance of affirming this principle across the board. 33

Malcolm Muggeridge, the English writer, goes right to the heart of the matter: "Either life is always and in all circumstances sacred, or intrinsically of no account; it is inconceivable that it should be in some cases the one, and in some the other." The sanctity of innocent human life is a principle that Congress should proclaim at every opportunity. 34

It is possible that the Supreme Court itself may overturn its abortion rulings. We need only recall that in *Brown* v. *Board of Education* the court reversed its own earlier "separate-but-equal" decision. I believe if the Supreme Court took another look at *Roe* v. *Wade*, and considered the real issue between the sanctity of life ethic and the quality of life ethic, it would change its mind once again. 35

As we continue to work to overturn *Roe* v. *Wade*, we must also continue to lay the groundwork for a society in which abortion is not the accepted answer to unwanted pregnancy. Pro-life people have already taken heroic steps, often at great personal sacrifice, to provide for unwed mothers. I recently spoke about a young pregnant woman named Victoria, who said, "In this society we save whales, we save timber wolves and bald eagles and Coke bottles. Yet, everyone wanted me to throw away my baby." She has been helped by Sav-a-Life, a group in Dallas, which provides a way for unwed mothers to preserve the human life within them when they might otherwise be tempted to resort to abortion. I think also of House of His Creation in Coatesville, Pennsylvania, where a loving couple has taken in almost two hundred young women in the past ten years. They have seen, as a fact of life, that the girls are *not* better off having abortions than saving their babies. I am also reminded of the remarkable Rossow family of Ellington, Connecticut, who have 36 opened their hearts and their home to nine handicapped adopted and foster children.

37   The Adolescent Family Life Program, adopted by Congress at the request of Senator Jeremiah Denton, has opened new opportunities for unwed mothers to give their children life. We should not rest until our entire society echoes the tone of John Powell in the dedication of his book, *Abortion: The Silent Holocaust,* a dedication to every woman carrying an unwanted child: "Please believe that you are not alone. There are many of us that truly love you, who want to stand at your side, and help in any way we can." And we can echo the always-practical woman of faith, Mother Teresa, when she says, "If you don't want the little child, that unborn child, give him to me." We have so many families in America seeking to adopt children that the slogan "every child a wanted child" is now the emptiest of all reasons to tolerate abortion.

38   I have often said we need to join in prayer to bring protection to the unborn. Prayer and action are needed to uphold the sanctity of human life. I believe it will not be possible to accomplish our work, the work of saving lives, "without being a soul of prayer." The famous British member of Parliament William Wilberforce prayed with his small group of influential friends, the "Clapham Sect," for *decades* to see an end to slavery in the British empire. Wilberforce led that struggle in Parliament, unflaggingly, because he believed in the sanctity of human life. He saw the fulfillment of his impossible dream when Parliament outlawed slavery just before his death.

39   Let his faith and perseverance be our guide. We will never recognize the true value of our own lives until we affirm the value in the life of others, a value of which Malcolm Muggeridge says: " . . . however low it flickers or fiercely burns, it is still a Divine flame which no man dare presume to put out, be his motives ever so humane and enlightened."

40   Abraham Lincoln recognized that we could not survive as a free land when some men could decide that others were not fit to be free and should therefore be slaves. Likewise, we cannot survive as a free nation when some men decide that others are not fit to live and should be abandoned to abortion or infanticide. My administration is dedicated to the preservation of America as a free land, and there is no cause more important for preserving that freedom than affirm-

ing the transcendent right to life of all human beings, the right
without which no other rights have any meaning.

STUDY QUESTIONS

1. Of the many arguments Reagan offers in this essay, what is
   his core argument for the conclusion that abortion is wrong?
2. At various places Reagan calls the fetus a "child," a "human
   life," and a "baby." What evidence does Reagan provide that
   these words are true indicators of a fetus's status?
3. Reagan connects abortion to the attitudes that lead to infanticide and slavery. Why does he think the "freedom of
   choice" position on abortion stems from these same attitudes?
   (Consider, for example, the "dreaded complication" that
   Reagan mentions in paragraph 29.)
4. In paragraphs 9 and 11, Reagan raises what he calls "the real
   issue" or "question" of abortion. Why does he think the
   question is *not* when human life begins?
5. How do you think Reagan would respond to the criteria of
   personhood presented in the selection by Mary Anne Warren
   (p. 131)?
6. Throughout the essay, Reagan quotes well-known individuals and authorities. Do these quotations supplement or replace argument?

Reference Chapters: 6, 7, 11

BARUCH BRODY

Fetal Humanity and Brain Function

Baruch Brody is Leon Jaworski Professor of Biomedical Ethics, Director of the Center for Ethics, Medicine, and Public Issues, and Professor of Medicine and Community Medicine at Baylor College of Medicine in Houston, Texas. In the following selection, Brody argues that while a fetus is not a human being from the point of conception, it becomes one early in the pregnancy. His strategy in this essay is unusual, since he reasons backwards from the conditions for death to a conclusion about when human life begins.

1   The question which we must now consider is the question of fetal humanity. Some have argued that the fetus is a human being with a right to life (or, for convenience, just a human being) from the moment of conception. Others have argued that the fetus only becomes a human being at the moment of birth. Many positions in between these two extremes have also been suggested. How are we to decide which is correct?

2   The analysis which we will propose here rests upon certain metaphysical assumptions which I have defended elsewhere. These assumptions are: (a) the question is when has the fetus acquired all the properties essential (necessary) for being a human being, for when it has, it is a human being; (b) these properties are such that

[Baruch Brody, "Fetal Humanity and the Theory of Essentialism," in Robert Baker and Frederick Elliston, eds., *Philosophy and Sex*. Buffalo, N.Y.: Prometheus Books, 1975, pp. 348–352.]

the loss of any one of them means that the human being in question has gone out of existence and not merely stopped being a human being; (c) human beings go out of existence when they die. It follows from these assumptions that the fetus becomes a human being when it acquires all those characteristics which are such that the loss of any one of them would result in the fetus's being dead. We must, therefore, turn to the analysis of death.

We will first consider the question of what properties are essen-   3
tial to being human if we suppose that death and the passing out of existence occur only if there has been an irreparable cessation of brain function (keeping in mind that that condition itself, as we have noted, is a matter of medical judgment). We shall then consider the same question on the supposition that [Paul] Ramsey's more complicated theory of death (the modified traditional view) is correct.

According to what is called the brain-death theory, as long as   4
there has not been an irreparable cessation of brain function the person in question continues to exist, no matter what else has happened to him. If so, it seems to follow that there is only one property—leaving aside those entailed by this one property—that is essential to humanity, namely, the possession of a brain that has not suffered an irreparable cessation of function.

Several consequences follow immediately from this conclusion.   5
We can see that a variety of often advanced claims about the essence of humanity are false. For example, the claim that movement, or perhaps just the ability to move, is essential for being human is false. A human being who has stopped moving, and even one who has lost the ability to move, has not therefore stopped existing. Being able to move, and a fortiori moving, are not essential properties of human beings and therefore are not essential to being human. Similarly, the claim that being perceivable by other human beings is essential for being human is also false. A human being who has stopped being perceivable by other humans (for example, someone isolated on the other side of the moon, out of reach even of radio communication) has not stopped existing. Being perceivable by other human beings is not an essential property of human beings and is not essential to being human. And the same point can be made about the claims that viability is essential for being human, that independent exist-

ence is essential for being human, and that actual interaction with other human beings is essential for being human. The loss of any of these properties would not mean that the human being in question had gone out of existence, so none of them can be essential to that human being and none of them can be essential for being human.

6 Let us now look at the following argument: (1) A functioning brain (or at least, a brain that, if not functioning, is susceptible of function) is a property that every human being must have because it is essential for being human. (2) By the time an entity acquires that property, it has all the other properties that are essential for being human. Therefore, when the fetus acquires that property it becomes a human being. It is clear that the property in question is, according to the brain-death theory, one that is had essentially by all human beings. The question that we have to consider is whether the second premise is true. It might appear that its truth does follow from the brain-death theory. After all, we did see that the theory entails that only one property (together with those entailed by it) is essential for being human. Nevertheless, rather than relying solely on my earlier argument, I shall adopt an alternative approach to strengthen the conviction that this second premise is true: I shall note the important ways in which the fetus resembles and differs from an ordinary human being by the time it definitely has a functioning brain (about the end of the sixth week of development). It shall then be evident, in light of our theory of essentialism, that none of these differences involves the lack of some property in the fetus that is essential for its being human.

7 Structurally, there are few features of the human being that are not fully present by the end of the sixth week. Not only are the familiar external features and all the internal organs present, but the contours of the body are nicely rounded. More important, the body is functioning. Not only is the brain functioning, but the heart is beating sturdily (the fetus by this time has its own completely developed vascular system), the stomach is producing digestive juices, the liver is manufacturing blood cells, the kidney is extracting uric acid from the blood, and the nerves and muscles are operating in concert, so that reflex reactions can begin.

8 What are the properties that a fetus acquires after the sixth week

of its development? Certain structures do appear later. These include the fingernails (which appear in the third month), the completed vocal chords (which also appear then), taste buds and salivary glands (again, in the third month), and hair and eyelashes (in the fifth month). In addition, certain functions begin later than the sixth week. The fetus begins to urinate (in the third month), to move spontaneously (in the third month), to respond to external stimuli (at least in the fifth month), and to breathe (in the sixth month). Moreover, there is a constant growth in size. And finally, at the time of birth the fetus ceases to receive its oxygen and food through the placenta and starts receiving them through the mouth and nose.

I will not examine each of these properties (structures and functions) to show that they are not essential for being human. The procedure would be essentially the one used previously to show that various essentialist claims are in error. We might, therefore, conclude, on the supposition that the brain-death theory is correct, that the fetus becomes a human being about the end of the sixth week after its development. 9

There is, however, one complication that should be noted here. There are, after all, progressive stages in the physical development and in the functioning of the brain. For example, the fetal brain (and nervous system) does not develop sufficiently to support spontaneous motion until some time in the third month after conception. There is, of course, no doubt that that stage of development is sufficient for the fetus to be human. No one would be likely to maintain that a spontaneously moving human being has died; and similarly, a spontaneously moving fetus would seem to have become human. One might, however, want to claim that the fetus does not become a human being until the point of spontaneous movement. So then, on the supposition that the brain-death theory of death is correct, one ought to conclude that the fetus becomes a human being at some time between the sixth and twelfth week after its conception. 10

But what if we reject the brain-death theory, and replace it with its equally plausible contender, Ramsey's theory of death? According to that theory—which we can call the brain, heart, and lung theory of death—the human being does not die, does not go out of 11 existence, until such time as the brain, heart and lungs have irreparably ceased functioning naturally. What are the essential features of being human according to this theory?

12   Actually, the adoption of Ramsey's theory requires no major modifications. According to that theory, what is essential to being human, what each human being must retain if he is to continue to exist, is the possession of a functioning (actually or potentially) heart, lung, or brain. It is only when a human being possesses none of these that he dies and goes out of existence; and the fetus comes into humanity, so to speak, when he acquires one of these.

13   On Ramsey's theory, the argument would now run as follows: (1) The property of having a functioning brain, heart, or lungs (or at least organs of the kind that, if not functioning, are susceptible of function) is one that every human being must have because it is essential for being human. (2) By the time that an entity acquires that property it has all the other properties that are essential for being human. Therefore, when the fetus acquires that property it becomes a human being. There remains, once more, the problem of the second premise. Since the fetal heart starts operating rather early, it is not clear that the second premise is correct. Many systems are not yet operating, and many structures are not yet present. Still, following our theory of essentialism, we should conclude that the fetus becomes a human being when it acquires a functioning heart (the first of the organs to function in the fetus).

14   There is, however, a further complication here, and it is analogous to the one encountered if we adopt the brain-death theory: When may we properly say that the fetal heart begins to function? At two weeks, when occasional contractions of the primitive fetal heart are present? In the fourth to fifth week, when the heart, although incomplete, is beating regularly and pumping blood cells through a closed vascular system, and when the tracings obtained by an ECG exhibit the classical elements of an adult tracing? Or after the end of the seventh week, when the fetal heart is functionally complete and "normal"?

15   We have not reached a precise conclusion in our study of the question of when the fetus becomes a human being. We do know that it does so some time between the end of the second week and the end of the third month. But it surely is not a human being at

the moment of conception and it surely is one by the end of the third month. Though we have not come to a final answer to our question, we have narrowed the range of acceptable answers considerably.

[In summary] we have argued that the fetus becomes a human being with a right to life some time between the second and twelfth week after conception. We have also argued that abortions are morally impermissible after that point except in rather unusual circumstances. What is crucial to note is that neither of these arguments appeal to any theological considerations. We conclude, therefore, that there is a human-rights basis for moral opposition to abortions.

16

STUDY QUESTIONS

1. The overall structure of Brody's argument is as follows:

   A human goes out of existence when it dies.
   A human dies when feature z is no longer present.

   Therefore, feature z is essential to being a human.
   Fetuses have feature z at time t.

   Therefore, fetuses are humans at time t.

   Therefore, abortions are immoral after time t.

   What premises are assumed between the final two lines of the overall argument?

2. Brody considers two theories about when death occurs: the brain-death theory and Ramsey's heart-lungs-brain theory. In paragraph 11, Brody states that the two theories are "equally plausible." On what evidence does he make this judgment?

3. In discussing Ramsey's theory of death (paragraphs 11 through 13), Brody makes an inference like the following:

   One is not dead unless one's heart, lungs, and brain are not functioning. (Paragraph 11)

   If one has a functioning heart, lungs, or brain, then one is alive. (Paragraph 13)

Let's symbolize the predicates as follows:

Hx = x has a functioning heart
Lx = x has functioning lungs
Bx = x has a functioning brain
Ax = x is alive

Does the following symbolization capture the form of this inference?

(x) $[\sim(\sim Hx \cdot \sim Lx \cdot \sim Bx) \supset Ax]$
(x) $[(Hx \vee Lx \vee Bx) \supset Ax]$

If not, what is the correct symbolization? Is the inference valid?

4. If we compare Brody's essay with Mary Anne Warren's "On the Moral and Legal Status of Abortion" (p. 131), we get two different accounts of the defining features of human beings. What is the fundamental point of disagreement that gives rise to the different definitions?

Reference Chapters: 3, 12, 13

LAURENCE TRIBE

Opposition to Abortion Is Not Based on Alleged Rights to Life

Laurence H. Tribe was born in Shanghai in 1941. He is Tyler Professor of Constitutional Law at Harvard Law School. In

[Laurence H. Tribe, excerpts from *Abortion: The Clash of Absolutes.* New York: W. W. Norton, 1990, pp. 231–234, references deleted.]

the following selection, excerpted from his book Abortion: The Clash of Absolutes, *Tribe offers an analysis of the premises underlying opposition to abortion.*

Most of those who regard abortion as, at best, a necessary evil would nonetheless make an exception permitting abortion in cases of rape and incest. (This exception is really about rape. Most cases of incest probably involve an older relative and a young child, and so in most people's experience, incest is really a particular kind of rape.) Although polling numbers on abortion are notoriously sensitive to the wording of the questions (the *New York Times* has observed that "one of every six Americans says simultaneously that abortion is murder and that it is sometimes the best course"), the polls do reveal this truth quite starkly.

One nationwide poll, for example, showed that 40 percent of the American public oppose abortion when it is sought because "the mother is an unmarried teenager whose future life might be seriously affected." Yet 81 percent favor abortion "if the woman became pregnant because of rape or incest." Only 17 percent oppose abortion in such cases. This suggests that almost 60 percent of those who oppose abortion for the unmarried teenager would support it in cases of rape or incest.

Regional polls corroborate the hypothesis that people who generally oppose abortion would nonetheless permit it in cases of rape and incest. For example, polls in Florida in the weeks after the *Webster* decision showed that although 59 percent of the registered voters said they agreed that during the first trimester of pregnancy the decision to have an abortion "should be left entirely to a woman and her doctor," 53 percent of those polled said that abortion should be "illegal" when sought because the woman's family "has a very low income and cannot afford to have any more children," and 60 percent said that abortion should not be permitted where sought because "the pregnancy would interfere with the mother's work or education." Still, 78 percent of those polled thought abortion should be available in cases where the pregnancy resulted from rape or incest. Only 13 percent said it should not. This suggests that more than 75 percent of the people who oppose abortion in circumstances of economic or personal hardship may well accept it in cases of rape and incest.

4   A similar poll in Utah found that, although 58 percent of the
adult population thought abortion should not be available "to
women who choose it in the first trimester" and 68 percent thought
it should not be available to women who choose it in the second
trimester, prior to viability, again an overwhelming majority (81
percent) agreed that abortion should be available in cases of rape
and incest. Only 11 percent disagreed. This suggests that up to 80
percent of Utah residents who oppose abortion on request in the first
trimester of pregnancy may nonetheless support the availability of
abortion in cases of rape and incest.

5   Support of a rape exception makes plain that most people's
opposition to abortion, unlike their opposition to murder, *can* be
overridden. It therefore suggests that antiabortion sentiment is not
entirely rooted in a belief that abortion constitutes the killing of an
innocent human being. It is hard to see how any such justification
for limiting abortion could plausibly be put forward by anyone who
thinks that abortion should be permitted in cases of rape. A fetus
conceived as a result of a violent rape is no less innocent than one
conceived in a mutually desired act of love. The fetus obviously is
not responsible for the circumstances surrounding its conception.
Yet the vast majority of people who oppose abortion would permit
such a fetus to be destroyed, even if they were rewriting from scratch
the constitutional rules governing this thorny topic.

6   If support for a rape and incest exception suggests that most
opposition to abortion is *not* entirely about the destruction of innocent human life it might also reveal something about the views,
conscious or unconscious, that lie at the heart of the belief that in
general, access to abortion should be restricted.

7   Surely there should be nothing abhorrent about the *particular*
fetuses that are the products of rapes. It is true that a position in
favor of denying criminals the right to reproduce has at times been
expressed in the United States—for example in the Oklahoma law
struck down in the 1940s by the Supreme Court in *Skinner v.
Oklahoma*, a law that provided for the sterilization of anyone previously found guilty two or more times of "felonies involving moral
turpitude." But a desire to deny the rapist his child could hardly
explain a willingness to make abortion available to women who
have been raped. And any notion that the fetus itself is tainted by a

kind of "original sin" seems most implausible. After all, when the woman who has been raped *chooses* to give birth to the rapist's child rather than to abort, she is commended, not condemned.

Right-to-life advocates who would allow abortion for a woman who becomes pregnant after a rape are probably reacting out of compassion for the woman; they don't think she should have to live through having her rapist's child develop within her. But the only thing to distinguish that from any other unwanted pregnancy is the nature of the sexual activity out of which the pregnancy arose. A fetus resulting from rape or (in most cases) from incest is the product of a sex act to which the woman did not consent. It is only the *nonconsensual nature of the sex* that led to her pregnancy that could make abortion in the case of rape seem justified to someone who would condemn all other abortions not needed to save the pregnant woman's life.

This in turn suggests that one's opposition to such *other* abortions reflects a sense that continued pregnancy is simply the price women must pay for engaging in *consensual sex*. The lack of sympathy toward women who have experienced contraceptive failure suggests that many of us have no discomfort at the idea that women who choose to have sex simply cannot be allowed to avoid *some* risk of a pregnancy that they will just have to carry to term.

As we have seen, this feeling may be partly rooted in the belief that in general, this is the way of nature. In this view, opposition to abortion may reflect an ambivalence about the use of technology in general, in this case medical technology, to overcome that which always before seemed "natural": that sex would lead to pregnancy and pregnancy to childbirth.

But notice how often this feeling about abortion is held by people who generally welcome the energetic uses of new technologies. Especially if such people regard nervousness about nuclear reactors or computers or other "unnatural" developments as silly or childish, their aversion to abortion rights would seem to reflect a deeply held *sexual* morality, in which pregnancy and childbirth are seen as a punishment that women in particular must endure for engaging in consensual sex. The fact that opposition to abortion rights may in large part be about sexual morality is reflected, too, in the attitude, noted earlier, of those who oppose abortion and seem willing to do almost anything to stop it—*except* take the effective pregnancy-reducing step of providing birth control education and better contraceptives.

12   At least to *these* "pro-life" activists, it seems to be more important to prevent the marginal increase in sexual activity that they believe will follow from sex education and the availability of birth control than to lower the number of abortions being performed. Theirs is thus a position in which sexual morality is primary, with any claim of a fetus's right to life taking a very distant backseat.

STUDY QUESTIONS

1. Tribe's core argument is as follows: If the opposition to abortion were based on the rights of the fetus, then opponents would not allow an exception for pregnancies resulting from rape; but they do allow an exception for rape; hence opposition to abortion is not based on the rights of the fetus. Identify the argument for each of the premises in Tribe's core argument.

2. Tribe cites data from a number of polls to support the conclusion that most people who generally oppose abortion would make exceptions in cases of rape. Are the polls he cites representative enough to support this conclusion?

3. What is Tribe's best explanation for why many of those opposed to abortion will allow exceptions in cases of rape?

4. What other explanations does Tribe consider and reject? Are they the only possible explanations?

5. Does Tribe say anything in this passage that would be a problem for those opposed to abortion in all cases?

6. Does Tribe prove his conclusion that much opposition to abortion is not based on the view that a fetus has a right to life?

Reference Chapters: 7, 10, 15, 18

MARY ANNE WARREN

On the Moral and Legal Status of Abortion

Mary Anne Warren, author of Gendercide *and* The Nature of Woman, *is a professor of philosophy at San Francisco State University in California. In the following selection, she defends a woman's right to abortion.*

The question which we must answer in order to produce a satisfactory solution to the problem of the moral status of abortion is this: How are we to define the moral community, the set of beings with full and equal moral rights, such that we can decide whether a human fetus is a member of this community or not? What sort of entity, exactly, has the inalienable rights to life, liberty, and the pursuit of happiness? Jefferson attributed these rights to all *men*, and it may or may not be fair to suggest that he intended to attribute them *only* to men. Perhaps he ought to have attributed them to all human beings. If so, then we arrive, first, at Noonan's problem of defining what makes a being human, and, second, at the equally vital question which Noonan does not consider, namely, What reason is there for identifying the moral community with the set of all human beings, in whatever way we have chosen to define that term?

[Mary Anne Warren, Part II of "On the Moral and Legal Status of Abortion." *The Monist* 57 (January 1973), pp. 52–61.]

1. On the Definition of 'Human'

2   One reason why this vital second question is so frequently overlooked in the debate over the moral status of abortion is that the
term 'human' has two distinct, but not often distinguished, senses.
This fact results in a slide of meaning, which serves to conceal the
fallaciousness of the traditional argument that since (1) it is wrong
to kill innocent human beings, and (2) fetuses are innocent human
beings, then (3) it is wrong to kill fetuses. For if 'human' is used in
the same sense in both (1) and (2) then, whichever of the two senses
is meant, one of these premises is question-begging. And if it is used
in two different senses then of course the conclusion doesn't follow.

3   Thus, (1) is a self-evident moral truth,[1] and avoids begging the
question about abortion, only if 'human being' is used to mean
something like "a full-fledged member of the moral community."
(It may or may not also be meant to refer exclusively to members
of the species *Homo sapiens.*) We may call this the *moral* sense of
'human.' It is not to be confused with what we will call the *genetic*
sense, i.e., the sense in which *any* member of the species is a human
being, and no member of any other species could be. If (1) is
acceptable only if the moral sense is intended, (2) is non-question-
begging only if what is intended is the genetic sense.

4   In "Deciding Who is Human," Noonan argues for the classification of fetuses with human beings by pointing to the presence of the
full genetic code, and the potential capacity for rational thought. It
is clear that what he needs to show, for his version of the traditional
argument to be valid, is that fetuses are human in the moral sense,
the sense in which it is analytically true that all human beings have
full moral rights. But, in the absence of any argument showing that
whatever is genetically human is also morally human, and he gives
none, nothing more than genetic humanity can be demonstrated by
the presence of the human genetic code. And, as we will see, the
potential capacity for rational thought can at most show that an
entity has the potential for *becoming* human in the moral sense.

---

1. Of course, the principle that it is (always) wrong to kill innocent human
beings is in need of many other modifications, e.g., that it may be permissible to
do so to save a greater number of other innocent human beings, but we may safely
ignore these complications here.

2. Defining the Moral Community

Can it be established that genetic humanity is sufficient for moral   5
humanity? I think that there are very good reasons for not defining
the moral community in this way. I would like to suggest an
alternative way of defining the moral community, which I will
argue for only to the extent of explaining why it is, or should be,
self-evident. The suggestion is simply that the moral community
consists of all and only *people,* rather than all and only human
beings;[2] and probably the best way of demonstrating its self-evi-
dence is by considering the concept of personhood, to see what sorts
of entity are and are not persons, and what the decision that a being
is or is not a person implies about its moral rights.

What characteristics entitle an entity to be considered a person?   6
This is obviously not the place to attempt a complete analysis of the
concept of personhood, but we do not need such a fully adequate
analysis just to determine whether and why a fetus is or isn't a
person. All we need is a rough and approximate list of the most basic
criteria of personhood, and some idea of which, or how many, of
these an entity must satisfy in order to properly be considered a
person.

In searching for such criteria, it is useful to look beyond the set   7
of people with whom we are acquainted, and ask how we would
decide whether a totally alien being was a person or not. (For we
have no right to assume that genetic humanity is necessary for
personhood.) Imagine a space traveler who lands on an unknown
planet and encounters a race of beings utterly unlike any he has ever
seen or heard of. If he wants to be sure of behaving morally toward
these beings, he has to somehow decide whether they are people,
and hence have full moral rights, or whether they are the sort of
thing which he need not feel guilty about treating as, for example,
a source of food.

How should he go about making this decision? If he has some   8
anthropological background, he might look for such things as religion, art, and the manufacturing of tools, weapons, or shelters, since

2. From here on, we will use 'human' to mean genetically human, since the
moral sense seems closely connected to, and perhaps derived from, the assumption
that genetic humanity is sufficient for membership in the moral community.

these factors have been used to distinguish our human from our prehuman ancestors, in what seems to be closer to the moral than the genetic sense of 'human.' And no doubt he would be right to consider the presence of such factors as good evidence that the alien beings were people, and morally human. It would, however, be overly anthropocentric of him to take the absence of these things as adequate evidence that they were not, since we can imagine people who have progressed beyond, or evolved without ever developing, these cultural characteristics.

9   I suggest that the traits which are most central to the concept of personhood, or humanity in the moral sense, are, very roughly, the following:

(1) consciousness (of objects and events external and/or internal to the being), and in particular the capacity to feel pain;
(2) reasoning (the *developed* capacity to solve new and relatively complex problems);
(3) self-motivated activity (activity which is relatively independent of either genetic or direct external control);
(4) the capacity to communicate, by whatever means, messages of an indefinite variety of types, that is, not just with an indefinite number of possible contents, but on indefinitely many possible topics;
(5) the presence of self-concepts, and self-awareness, either individual or racial, or both.

10   Admittedly, there are apt to be a great many problems involved in formulating precise definitions of these criteria, let alone in developing universally valid behavioral criteria for deciding when they apply. But I will assume that both we and our explorer know approximately what (1)–(5) mean, and that he is also able to determine whether or not they apply. How, then, should he use his findings to decide whether or not the alien beings are people? We needn't suppose that an entity must have *all* of these attributes to be properly considered a person; (1) and (2) alone may well be sufficient for personhood, and quite probably (1)–(3) are sufficient. Neither do we need to insist that any one of these criteria is *necessary* for personhood, although once again (1) and (2) look like fairly good

candidates for necessary conditions, as does (3), if 'activity' is construed so as to include the activity of reasoning.

All we need to claim, to demonstrate that a fetus is not a person, [11] is that any being which satisfies *none* of (1)–(5) is certainly not a person. I consider this claim to be so obvious that I think anyone who denied it, and claimed that a being which satisfied none of (1)–(5) was a person all the same, would thereby demonstrate that he had no notion at all of what a person is—perhaps because he had confused the concept of a person with that of genetic humanity. If the opponents of abortion were to deny the appropriateness of these five criteria, I do not know what further arguments would convince them. We would probably have to admit that our conceptual schemes were indeed irreconcilably different, and that our dispute could not be settled objectively.

I do not expect this to happen, however, since I think that the [12] concept of a person is one which is very nearly universal (to people), and that it is common to both proabortionists and antiabortionists, even though neither group has fully realized the relevance of this concept to the resolution of their dispute. Furthermore, I think that on reflection even the antiabortionists ought to agree not only that (1)–(5) are central to the concept of personhood, but also that it is a part of this concept that all and only people have full moral rights. The concept of a person is in part a moral concept; once we have admitted that x is a person we have recognized, even if we have not agreed to respect, x's right to be treated as a member of the moral community. It is true that the claim that x is a *human being* is more commonly voiced as part of an appeal to treat x decently than is the claim that x is a person, but this is either because 'human being' is here used in the sense which implies personhood, or because the genetic and moral senses of 'human' have been confused.

Now if (1)–(5) are indeed the primary criteria of personhood, [13] then it is clear that genetic humanity is neither necessary nor sufficient for establishing that an entity is a person. Some human beings are not people, and there may well be people who are not human beings. A man or woman whose consciousness has been permanently obliterated but who remains alive is a human being which is no longer a person; defective human beings, with no appreciable mental capacity, are not and presumably never will be people; and a fetus is a human being which is not yet a person, and which therefore cannot coherently be said to have full moral rights. Citizens of the next century should be prepared to recognize highly advanced, self-aware robots or computers, should such be developed, and intelligent inhabitants of other worlds, should such be found, as people in the fullest sense, and to respect their moral rights. But to ascribe full moral rights to an entity which is not a person is as absurd as to ascribe moral obligations and responsibilities to such an entity.

3. Fetal Development and the Right to Life

14   Two problems arise in the application of these suggestions for the definition of the moral community to the determination of the precise moral status of a human fetus. Given that the paradigm example of a person is a normal adult human being, then (1) How like this paradigm, in particular how far advanced since conception, does a human being need to be before it begins to have a right to life by virtue, not of being fully a person as of yet, but of being *like* a person? and (2) To what extent, if any, does the fact that a fetus has the *potential* for becoming a person endow it with some of the same rights? Each of these questions requires some comment.

15   In answering the first question, we need not attempt a detailed consideration of the moral rights of organisms which are not developed enough, aware enough, intelligent enough, etc., to be considered people, but which resemble people in some respects. It does seem reasonable to suggest that the more like a person, in the relevant respects, a being is, the stronger is the case for regarding it as having a right to life, and indeed the stronger its right to life is. Thus we ought to take seriously the suggestion that, insofar as "the human individual develops biologically in a continuous fashion . . . the rights of a human person might develop in the same way."[3] But

3. Thomas L. Hayes, "A Biological View," *Commonweal*, 85 (March 17, 1967), 677–78; quoted by Daniel Callahan, in *Abortion, Law, Choice, and Morality* (London: Macmillan & Co., 1970).

we must keep in mind that the attributes which are relevant in determining whether or not an entity is enough like a person to be regarded as having some of the same moral rights are no different from those which are relevant to determining whether or not it is fully a person—i.e., are no different from (1)–(5)—and that being genetically human, or having recognizably human facial and other physical features, or detectable brain activity, or the capacity to survive outside the uterus, are simply not among these relevant attributes.

Thus it is clear that even though a seven- or eight-month fetus   16
has features which make it apt to arouse in us almost the same powerful protective instinct as is commonly aroused by a small infant, nevertheless it is not significantly more personlike than is a very small embryo. It is *somewhat* more personlike; it can apparently feel and respond to pain, and it may even have a rudimentary form of consciousness, insofar as its brain is quite active. Nevertheless, it seems safe to say that it is not fully conscious, in the way that an infant of a few months is, and that it cannot reason, or communicate messages of indefinitely many sorts, does not engage in self-motivated activity, and has no self-awareness. Thus, in the *relevant* respects, a fetus, even a fully developed one, is considerably less personlike than is the average mature mammal, indeed the average fish. And I think that a rational person must conclude that if the right to life of a fetus is to be based upon its resemblance to a person, then it cannot be said to have any more right to life than, let us say, a newborn guppy (which also seems to be capable of feeling pain), and that a right of that magnitude could never override a woman's right to obtain an abortion, at any stage of her pregnancy.

There may, of course, be other arguments in favor of placing legal   17
limits upon the stage of pregnancy in which an abortion may be performed. Given the relative safety of the new techniques of artifically inducing labor during the third trimester, the danger to the woman's life or health is no longer such an argument. Neither is the fact that people tend to respond to the thought of abortion in the later stages of pregnancy with emotional repulsion, since mere emotional responses cannot take the place of moral reasoning in determining what ought to be permitted. Nor, finally, is the frequently heard argument that legalizing abortion, especially late in

the pregnancy, may erode the level of respect for human life, leading, perhaps, to an increase in unjustified euthanasia and other crimes. For this threat, if it is a threat, can be better met by educating people to the kinds of moral distinctions which we are making here than by limiting access to abortion (which limitation may, in its disregard for the rights of women, be just as damaging to the level of respect for human rights).

18   Thus, since the fact that even a fully developed fetus is not personlike enough to have any significant right to life on the basis of its personlikeness shows that no legal restrictions upon the stage of pregnancy in which an abortion may be performed can be justified on the grounds that we should protect the rights of the older fetus; and since there is no other apparent justification for such restrictions, we may conclude that they are entirely unjustified. Whether or not it would be *indecent* (whatever that means) for a woman in her seventh month to obtain an abortion just to avoid having to postpone a trip to Europe, it would not, in itself, be *immoral,* and therefore it ought to be permitted.

4. Potential Personhood and the Right to Life

19   We have seen that a fetus does not resemble a person in any way which can support the claim that it has even some of the same rights. But what about its *potential,* the fact that if nurtured and allowed to develop naturally it will very probably become a person? Doesn't that alone give it at least some right to life? It is hard to deny that the fact that an entity is a potential person is a strong prima facie reason for not destroying it; but we need not conclude from this that a potential person has a right to life, by virtue of that potential. It may be that our feeling that it is better, other things being equal, not to destroy a potential person is better explained by the fact that potential people are still (felt to be) an invaluable resource, not to be lightly squandered. Surely, if every speck of dust were a potential person, we would be much less apt to conclude that every potential person has a right to become actual.

Still, we do not need to insist that a potential person has no right   20
to life whatever. There may well be something immoral, and not
just imprudent, about wantonly destroying potential people, when
doing so isn't necessary to protect anyone's rights. But even if a
potential person does have some prima facie right to life, such a
right could not possibly outweigh the right of a woman to obtain an
abortion, since the rights of any actual person invariably outweigh
those of any potential person, whenever the two conflict. Since this
may not be immediately obvious in the case of a human fetus, let
us look at another case.

Suppose that our space explorer falls into the hands of an alien   21
culture, whose scientists decide to create a few hundred thousand or
more human beings, by breaking his body into its component cells,
and using these to create fully developed human beings, with, of
course, his genetic code. We may imagine that each of these newly
created men will have all of the original man's abilities, skills,
knowledge, and so on, and also have an individual self-concept, in
short that each of them will be a bona fide (though hardly unique)
person. Imagine that the whole project will take only seconds, and
that its chances of success are extremely high, and that our explorer
knows all of this, and also knows that these people will be treated
fairly. I maintain that in such a situation he would have every right
to escape if he could, and thus to deprive all of these potential people
of their potential lives; for his right to life outweighs all of theirs
together, in spite of the fact that they are all genetically human, all
innocent, and all have a very high probability of becoming people
very soon, if only he refrains from acting.

Indeed, I think he would have a right to escape even if it were   22
not his life which the alien scientists planned to take, but only a year
of his freedom, or, indeed, only a day. Nor would he be obligated to
stay if he had gotten captured (thus bringing all these people-po-
tentials into existence) because of his own carelessness, or even if he
had done so deliberately, knowing the consequences. Regardless of
how he got captured, he is not morally obligated to remain in
captivity for *any* period of time for the sake of permitting any
number of potential people to come into actuality, so great is the
margin by which one actual person's right to liberty outweighs
whatever right to life even a hundred thousand potential people
have. And it seems reasonable to conclude that the rights of a

woman will outweigh by a similar margin whatever right to life a fetus may have by virtue of its potential personhood.

23   Thus, neither a fetus's resemblance to a person, nor its potential for becoming a person provides any basis whatever for the claim that it has any significant right to life. Consequently, a woman's right to protect her health, happiness, freedom, and even her life,[4] by terminating an unwanted pregnancy, will always override whatever right to life it may be appropriate to ascribe to a fetus, even a fully developed one. And thus, in the absence of any overwhelming social need for every possible child, the laws which restrict the right to obtain an abortion, or limit the period of pregnancy during which an abortion may be performed, are a wholly unjustified violation of a woman's most basic moral and constitutional rights.[5]

STUDY QUESTIONS

1. In paragraph 3, Warren says that 'human' has two senses. What are they? And what is the distinction Warren makes in paragraph 5 between people and human beings?

2. What is the point of the thought experiment in paragraphs 7 and 8 involving the space traveler who encounters aliens on an unknown planet? Does Warren uses it as part of her argument?

3. In paragraph 15, Warren rejects four attributes as candidates for determining that a fetus has rights. What are they, and on what grounds does she reject them?

4. In paragraph 16, Warren argues that even if a fetus has some degree of a right to life, "a right of that magnitude could never override a woman's right to obtain an abortion, at any stage of her pregnancy." Has she yet argued that a woman has a right to obtain an abortion? Where does she present her reasons justifying a woman's right to abortion?

4. That is, insofar as the death rate, for the woman, is higher for childbirth than for early abortion.

5. My thanks to the following people, who were kind enough to read and criticize an earlier version of this paper: Herbert Gold, Gene Glass, Anne Lauterbach, Judith Thomson, Mary Mothersill, and Timothy Binkley.

5. In the selection by Ronald Reagan (p. 109), abortion is linked to a lack of respect for human life and thus to crimes such as infanticide and slavery. Does Warren respond to the following objection: "Since you think that a woman may have an abortion at any point during her pregnancy, and since a fetus one day before birth and an infant one day after birth are not significantly different in their attributes, your position logically implies that infanticide is acceptable"?

6. What is the point of the example in paragraph 21 involving the space explorer who falls into the hands of aliens who wish to clone him?

Reference Chapters: 3, 7, 11

PART FIVE

On Dinosaurs, Perception, and Table Manners

STEPHEN JAY GOULD

Sex, Drugs, Disasters, and the Extinction of Dinosaurs

Stephen Jay Gould, born in New York City in 1941, is Professor of Geology at Harvard University. He is the author of numerous popular essays on science, many of which have been collected into books such as The Panda's Thumb, Hen's Teeth and Horses' Toes, *and* The Flamingo's Smile, *from which the following essay is taken.*

Science, in its most fundamental definition, is a fruitful mode of inquiry, not a list of enticing conclusions. The conclusions are the consequence, not the essence.

1

[Stephen J. Gould, "Sex, Drugs, Disasters, and the Extinction of Dinosaurs," in *The Flamingo's Smile: Reflections in Natural History.* New York: W. W. Norton, 1985, pp. 417–426.]

2   My greatest unhappiness with most popular presentations of science concerns their failure to separate fascinating claims from the methods that scientists use to establish the facts of nature. Journalists, and the public, thrive on controversial and stunning statements. But science is, basically, a way of knowing—in P. B. Medawar's apt words, "the art of the soluble." If the growing corps of popular science writers would focus on *how* scientists develop and defend those fascinating claims, they would make their greatest possible contribution to public understanding.

3   Consider three ideas, proposed in perfect seriousness to explain that greatest of all titillating puzzles—the extinction of dinosaurs. Since these three notions invoke the primally fascinating themes of our culture—sex, drugs, and violence—they surely reside in the category of fascinating claims. I want to show why two of them rank as silly speculation, while the other represents science at its grandest and most useful.

4   Science works with testable proposals. If, after much compilation and scrutiny of data, new information continues to affirm a hypothesis, we may accept it provisionally and gain confidence as further evidence mounts. We can never be completely sure that a hypothesis is right, though we may be able to show with confidence that it is wrong. The best scientific hypotheses are also generous and expansive: they suggest extensions and implications that enlighten related, and even far distant, subjects. Simply consider how the idea of evolution has influenced virtually every intellectual field.

5   Useless speculation, on the other hand, is restrictive. It generates no testable hypothesis, and offers no way to obtain potentially refuting evidence. Please note that I am not speaking of truth or falsity. The speculation may well be true; still, if it provides, in principle, no material for affirmation or rejection, we can make nothing of it. It must simply stand forever as an intriguing idea. Useless speculation turns in on itself and leads nowhere; good science, containing both seeds for its potential refutation and implications for more and different testable knowledge, reaches out. But, enough preaching. Let's move on to dinosaurs, and the three proposals for their extinction.

    1. Sex: Testes function only in a narrow range of temperature (those of mammals hang externally in a scrotal sac because internal body

temperatures are too high for their proper function). A worldwide rise in temperature at the close of the Cretaceous period caused the testes of dinosaurs to stop functioning and led to their extinction by sterilization of males.

2. Drugs: Angiosperms (flowering plants) first evolved toward the end of the dinosaurs' reign. Many of these plants contain psychoactive agents, avoided by mammals today as a result of their bitter taste. Dinosaurs had neither means to taste the bitterness nor livers effective enough to detoxify the substances. They died of massive overdoses.

3. Disasters: A large comet or asteroid struck the earth some 65 million years ago, lofting a cloud of dust into the sky and blocking sunlight, thereby suppressing photosynthesis and so drastically lowering world temperatures that dinosaurs and hosts of other creatures became extinct.

Before analyzing these three tantalizing statements, we must establish a basic ground rule often violated in proposals for the dinosaurs' demise. *There is no separate problem of the extinction of dinosaurs.* Too often we divorce specific events from their wider contexts and systems of cause and effect. The fundamental fact of dinosaur extinction is its synchrony with the demise of so many other groups across a wide range of habitats, from terrestrial to marine.

The history of life has been punctuated by brief episodes of mass 6 extinction. A recent analysis by University of Chicago paleontologists Jack Sepkoski and Dave Raup, based on the best and most exhaustive tabulation of data ever assembled, shows clearly that five episodes of mass dying stand well above the "background" extinctions of normal times (when we consider all mass extinctions, large and small, they seem to fall in a regular 26-million-year cycle). The Cretaceous debacle, occurring 65 million years ago and separating the Mesozoic and Cenozoic eras of our geological time scale, ranks prominently among the five. Nearly all the marine plankton (single-celled floating creatures) died with geological suddenness; among marine invertebrates, nearly 15 percent of all families perished, including many previously dominant groups, especially the ammonites (relatives of squids in coiled shells). On land, the dinosaurs disappeared after more than 100 million years of unchallenged domination.

In this context, speculations limited to dinosaurs alone ignore the 7 larger phenomenon. We need a coordinated explanation for a system of events that includes the extinction of dinosaurs as one component. Thus it makes little sense, though it may fuel our desire to view mammals as inevitable inheritors of the earth, to guess that dinosaurs died because small mammals ate their eggs (a perennial favorite among untestable speculations). It seems most unlikely that some disaster peculiar to dinosaurs befell these massive beasts—and that the debacle happened to strike just when one of history's five great dyings had enveloped the earth for completely different reasons.

8   The testicular theory, an old favorite from the 1940s, had its root in an interesting and thoroughly respectable study of temperature tolerances in the American alligator, published in the staid *Bulletin of the American Museum of Natural History* in 1946 by three experts on living and fossil reptiles—E. H. Colbert, my own first teacher in paleontology; R. B. Cowles; and C. M. Bogert.

9   The first sentence of their summary reveals a purpose beyond alligators: "This report describes an attempt to infer the reactions of extinct reptiles, especially the dinosaurs, to high temperatures as based upon reactions observed in the modern alligator." They studied, by rectal thermometry, the body temperatures of alligators under changing conditions of heating and cooling. (Well, let's face it, you wouldn't want to try sticking a thermometer under a 'gator's tongue.) The predictions under test go way back to an old theory first stated by Galileo in the 1630s—the unequal scaling of surfaces and volumes. As an animal, or any object, grows (provided its shape doesn't change), surface areas must increase more slowly than volumes—since surfaces get larger as length squared, while volumes increase much more rapidly, as length cubed. Therefore, small animals have high ratios of surface to volume, while large animals cover themselves with relatively little surface.

10   Among cold-blooded animals lacking any physiological mechanism for keeping their temperatures constant, small creatures have a hell of a time keeping warm—because they lose so much heat through their relatively large surfaces. On the other hand, large animals, with their relatively small surfaces, may lose heat so slowly that, once warm, they may maintain effectively constant temperatures against ordinary fluctuations of climate. (In fact, the resolution of the "hot-blooded dinosaur" controversy that burned so

brightly a few years back may simply be that, while large dinosaurs possessed no physiological mechanism for constant temperature, and were not therefore warm-blooded in the technical sense, their large size and relatively small surface area kept them warm.)

Colbert, Cowles, and Bogert compared the warming rates of small   11
and large alligators. As predicted, the small fellows heated up (and cooled down) more quickly. When exposed to a warm sun, a tiny 50-gram (1.76-ounce) alligator heated up one degree Celsius every minute and a half, while a large alligator, 260 times bigger at 13,000 grams (28.7 pounds), took seven and a half minutes to gain a degree. Extrapolating up to an adult 10-ton dinosaur, they concluded that a one-degree rise in body temperature would take eighty-six hours. If large animals absorb heat so slowly (through their relatively small surfaces), they will also be unable to shed any excess heat gained when temperatures rise above a favorable level.

The authors then guessed that large dinosaurs lived at or near   12
their optimum temperatures; Cowles suggested that a rise in global temperatures just before the Cretaceous extinction caused the dinosaurs to heat up beyond their optimal tolerance—and, being so large, they couldn't shed the unwanted heat. (In a most unusual statement within a scientific paper, Colbert and Bogert then explicitly disavowed this speculative extension of their empirical work on alligators.) Cowles conceded that this excess heat probably wasn't enough to kill or even to enervate the great beasts, but since testes often function only within a narrow range of temperature, he proposed that this global rise might have sterilized all the males, causing extinction by natural contraception.

The overdose theory has recently been supported by UCLA   13
psychiatrist Ronald K. Siegel. Siegel has gathered, he claims, more than 2,000 records of animals who, when given access, administer various drugs to themselves—from a mere swig of alcohol to massive doses of the big H. Elephants will swill the equivalent of twenty beers at a time, but do not like alcohol in concentrations greater than 7 percent. In a silly bit of anthropocentric speculation, Siegel states that "elephants drink, perhaps, to forget . . . the anxiety produced by shrinking rangeland and the competition for food."

Since fertile imaginations can apply almost any hot idea to the   14
extinction of dinosaurs, Siegel found a way. Flowering plants did not evolve until late in the dinosaurs' reign. These plants also

produced an array of aromatic, amino-acid-based alkaloids—the major group of psychoactive agents. Most mammals are "smart" enough to avoid these potential poisons. The alkaloids simply don't taste good (they are bitter); in any case, we mammals have livers happily supplied with the capacity to detoxify them. But, Siegel speculates, perhaps dinosaurs could neither taste the bitterness nor detoxify the substances once ingested. He recently told members of the American Psychological Association: "I'm not suggesting that all dinosaurs OD'd on plant drugs, but it certainly was a factor." He also argued that death by overdose may help explain why so many dinosaur fossils are found in contorted positions. (Do not go gentle into that good night.)[1]

15   Extraterrestrial catastrophes have long pedigrees in the popular literature of extinction, but the subject exploded again in 1979, after a long lull, when the father-son, physicist-geologist team of Luis and Walter Alvarez proposed that an asteroid, some 10 km in diameter, struck the earth 65 million years ago (comets, rather than asteroids, have since gained favor. Good science is self-corrective).

16   The force of such a collision would be immense, greater by far than the megatonnage of all the world's nuclear weapons. In trying to reconstruct a scenario that would explain the simultaneous dying of dinosaurs on land and so many creatures in the sea, the Alvarezes proposed that a gigantic dust cloud, generated by particles blown aloft in the impact, would so darken the earth that photosynthesis would cease and temperatures drop precipitously. (Rage, rage against the dying of the light.)[2] The single-celled photosynthetic oceanic plankton, with life cycles measured in weeks, would perish outright, but land plants might survive through the dormancy of their seeds (land plants were not much affected by the Cretaceous extinction, and any adequate theory must account for the curious pattern of differential survival). Dinosaurs would die by starvation and freezing; small, warm-blooded mammals, with more modest requirements for food and better regulation of body temperature, would squeak through. "Let the bastards freeze in the dark," as bumper stickers of our chauvinistic neighbors in sunbelt states

1. A reference to a poem of this title by the Welsh poet Dylan Thomas (1914–1953).—Eds.
2. A reference to the same poem.—Eds.

proclaimed several years ago during the Northeast's winter oil crisis.

All three theories, testicular malfunction, psychoactive overdos-   17
ing, and asteroidal zapping, grab our attention mightily. As pure
phenomenology, they rank about equally high on any hit parade of
primal fascination. Yet one represents expansive science, the others
restrictive and untestable speculation. The proper criterion lies in
evidence and methodology; we must probe behind the superficial
fascination of particular claims.

How could we possibly decide whether the hypothesis of testicu-   18
lar frying is right or wrong? We would have to know things that the
fossil record cannot provide. What temperatures were optimal for
dinosaurs? Could they avoid the absorption of excess heat by staying
in the shade, or in caves? At what temperatures did their testicles
cease to function? Were late Cretaceous climates ever warm enough
to drive the internal temperatures of dinosaurs close to this ceiling?
Testicles simply don't fossilize, and how could we infer their temperature tolerances even if they did? In short, Cowles's hypothesis
is only an intriguing speculation leading nowhere. The most damning statement against it appeared right in the conclusion of Colbert,
Cowles, and Bogert's paper, when they admitted: "It is difficult to
advance any definite arguments against this hypothesis." My statement may seem paradoxical—isn't a hypothesis really good if you
can't devise any arguments against it? Quite the contrary. It is
simply untestable and unusable.

Siegel's overdosing has even less going for it. At least Cowles   19
extrapolated his conclusion from some good data on alligators. And
he didn't completely violate the primary guideline of siting dinosaur
extinction in the context of a general mass dying—for rise in
temperature could be the root cause of a general catastrophe, zapping
dinosaurs by testicular malfunction and different groups for other
reasons. But Siegel's speculation cannot touch the extinction of
ammonites or oceanic plankton (diatoms make their own food with
good sweet sunlight; they don't OD on the chemicals of terrestrial
plants). It is simply a gratuitous, attention-grabbing guess. It cannot
be tested, for how can we know what dinosaurs tasted and what their
livers could do? Livers don't fossilize any better than testicles.

The hypothesis doesn't even make any sense in its own context.   20
Angiosperms were in full flower ten million years before dinosaurs

went the way of all flesh. Why did it take so long? As for the pains of a chemical death recorded in contortions of fossils, I regret to say (or rather I'm pleased to note for the dinosaurs' sake) that Siegel's knowledge of geology must be a bit deficient: muscles contract after death and geological strata rise and fall with motions of the earth's crust after burial—more than enough reason to distort a fossil's pristine appearance.

21   The impact story, on the other hand, has a sound basis in evidence. It can be tested, extended, refined and, if wrong, disproved. The Alvarezes did not just construct an arresting guess for public consumption. They proposed their hypothesis after laborious geochemical studies with Frank Asaro and Helen Michael had revealed a massive increase of iridium in rocks deposited right at the time of extinction. Iridium, a rare metal of the platinum group, is virtually absent from indigenous rocks of the earth's crust; most of our iridium arrives on extraterrestrial objects that strike the earth.

22   The Alvarez hypothesis bore immediate fruit. Based originally on evidence from two European localities, it led geochemists throughout the world to examine other sediments of the same age. They found abnormally high amounts of iridium everywhere—from continental rocks of the western United States to deep sea cores from the South Atlantic.

23   Cowles proposed his testicular hypothesis in the mid-1940s. Where has it gone since then? Absolutely nowhere, because scientists can do nothing with it. The hypothesis must stand as a curious appendage to a solid study of alligators. Siegel's overdose scenario will also win a few press notices and fade into oblivion. The Alvarezes' asteroid falls into a different category altogether, and much of the popular commentary has missed this essential distinction by focusing on the impact and its attendant results, and forgetting what really matters to a scientist—the iridium. If you talk just about asteroids, dust, and darkness, you tell stories no better and no more entertaining than fried testicles or terminal trips. It is the iridium—the source of testable evidence—that counts and forges the crucial distinction between speculation and science.

24   The proof, to twist a phrase, lies in the doing. Cowles's hypothesis has generated nothing in thirty-five years. Since its proposal in 1979, the Alvarez hypothesis has spawned hundreds of studies, a major conference, and attendant publications. Geologists are fired

up. They are looking for iridium at all other extinction boundaries. Every week exposes a new wrinkle in the scientific press. Further evidence that the Cretaceous iridium represents extraterrestrial impact and not indigenous volcanism continues to accumulate. As I revise this essay in November 1984 (this paragraph will be out of date when the book is published), new data include chemical "signatures" of other isotopes indicating unearthly provenance, glass spherules of a size and sort produced by impact and not by volcanic eruptions, and high-pressure varieties of silica formed (so far as we know) only under the tremendous shock of impact.

My point is simply this: Whatever the eventual outcome (I suspect it will be positive), the Alvarez hypothesis is exciting, fruitful science because it generates tests, provides us with things to do, and expands outward. We are having fun, battling back and forth, moving toward a resolution, and extending the hypothesis beyond its original scope.

As just one example of the unexpected, distant cross-fertilization that good science engenders, the Alvarez hypothesis made a major contribution to a theme that has riveted public attention in the past few months—so-called nuclear winter. In a speech delivered in April 1982, Luis Alvarez calculated the energy that a ten-kilometer asteroid would release on impact. He compared such an explosion with a full nuclear exchange and implied that all-out atomic war might unleash similar consequences.

This theme of impact leading to massive dust clouds and falling temperatures formed an important input to the decision of Carl Sagan and a group of colleagues to model the climatic consequences of nuclear holocaust. Full nuclear exchange would probably generate the same kind of dust cloud and darkening that may have wiped out the dinosaurs. Temperatures would drop precipitously and agriculture might become impossible. Avoidance of nuclear war is fundamentally an ethical and political imperative, but we must know the factual consequences to make firm judgments. I am heartened by a final link across disciplines and deep concerns—another criterion, by the way, of science at its best:[1] A recognition of

---

1. This quirky connection so tickles my fancy that I break my own strict rule about eliminating redundancies from these essays and end both this and the next piece with this prod to thought and action.

the very phenomenon that made our evolution possible by exterminating the previously dominant dinosaurs and clearing a way for the evolution of large mammals, including us, might actually help to save us from joining those magnificent beasts in contorted poses among the strata of the earth.

STUDY QUESTIONS

1. What definition of "science" does Gould propose?
2. On what grounds does Gould contrast good scientific hypotheses with useless speculations?
3. In paragraph 7, why is the hypothesis that small animals ate the dinosaurs' eggs dismissed as an untestable speculation?
4. Why are paragraphs 5 through 7 important to Gould's later evaluation of the proposed explanations of the extinction of the dinosaurs?
5. In paragraph 18, Gould argues that if you can't find any definite arguments against a given hypothesis, then the hypothesis has no value. Why does he believe this to be the case?
6. Reconstruct the "Sex" hypothesis and the "Drugs" hypothesis. In order to explain the extinction of the dinosaurs, what propositions do they depend upon? Is Gould right that none of those propositions is testable?
7. In what ways is the "Disaster" hypothesis both more testable and a more comprehensive explanation of the extinction of the dinosaurs?
8. In the end, how should we view the "Sex" and "Drugs" hypotheses? After all, couldn't they be true, even though they're not testable?

Reference Chapter: 18

MURDOCK PENCIL

Salt Passage Research: The State of the Art

Murdock Pencil, Professor of Social Darwinism at the Old School for Social Science Research, is the pseudonym of Michael Paconowsky, of the Institute for Communication Research at Stanford University. In the following selection, Paconowsky parodies social science research.

Conclusive evidence on the effects of the utterance "Please pass the salt" is found to be sadly lacking.

Strongly rooted in the English speech community is the belief that the utterance, "Please pass the salt," is efficacious in causing salt to move from one end of a table to the source of the utterance. In his *Canterbury Tales*, Chaucer notes:

> Shee I askked
> The salde to passe.
> Ne surprised was I
> Tha shee didde (4. p. 318).[1]

Similarly, Dickens writes:

> Old Heep did not become disgruntled at my obstinence. "Please pass the salt, Davey," he repeated coldly. I vacillated for a moment longer. Then I passed the salt, just as he knew I would (5. p. 278).

[Murdock Pencil, "Salt Passage Research: The State of the Art." *Journal of Communication* (Autumn 1976), pp. 31–36.]

1.   I asked her to pass the salt.
                         I was not surprised that she did.

2   The question of whether the movement of salt is causally dependent on the utterance of the phrase, "Please pass the salt," has occupied the attention of numerous philosophers (3, 9, 20). Empirical resolution of the validity of this belief, however, was not undertaken until the classic work of Hovland, Lumsdaine, and Sheffield (8) on the American soldier. Since then, numerous social scientists have explored the antecedent conditions that give rise to this apparent regularity. In this article, we will summarize those efforts that shed some light on the complex phenomenon known as salt passage.

3   Many social observers have noticed the apparent regularity with which salt travels from one end of the table to the source of the utterance "Please pass the salt." Hovland, Lumsdaine, and Sheffield (8), however, were the first to demonstrate empirically that the salt passage phenomenon was mediated by the presence of other people at the table. In a comparison of "others present" with "no others present" conditions, they found that when there were other people present at the table, there was a greater likelihood that the utterance, "Please pass the salt," would result in salt movement toward the source of the utterance. When there were *no other people* at the table, the utterance, "Please pass the salt," had no apparent effect. To test the possibility of a time delay involved in the "no others present" condition, Hovland, *et al.* arranged for 112 Army recruits, each sitting alone at one end of a table with salt at the other end, to repeat the utterance, "Please pass the salt," every five minutes for 12 hours. The average distance the salt traveled was .5 inch, which the experimenters explained was due to measurement error. The result of these two studies was, therefore, to demonstrate the importance of the presence of other people in the salt passage phenomenon.

4   Once the presence of other people was established as a necessary condition for salt passage as a consistent response to the utterance, "Please pass the salt," researchers began focusing on source and receiver characteristics that would affect salt passing behavior. Osgood and Tannenbaum (16) predicted greater compliance with salt passage utterances by high credible sources than by low credible sources. Newcomb (14) predicted greater compliance with sources who were perceived to have similar, rather than dissimilar, attitudes. Rokeach (17) predicted greater compliance for low dogmatic, rather than high dogmatic, people. McClelland (12) predicted

greater compliance for high N achievers than low N achievers. Surprisingly, no significant differences were found along any of these dimensions. Differences due to race were found, however, in the original Hovland, *et al.* study (8). Black soldiers were more likely to pass the salt to white soldiers, while white soldiers were less likely to pass the salt to black soldiers.[2]

Because source and receiver characteristics seemed to have little effect on the extent of salt passage, research attention turned its focus to the effects of message variables as the causal mechanism underlying this phenomenon.

Janis and Feshbach (10) found that other utterances were just as effective as "Please pass the salt" in achieving salt passage compliance. No significant differences in the extent of compliance were found due to the utterances, "Please pass the salt," "Would you mind passing the salt?" "Could I have the salt down here, buddy?" and "Salt!" Janis and Feshbach noted that in every successful utterance, the word "salt" was found. They concluded that the frequency of the sound waves associated with the phonemes in "salt" was in fact the causal mechanism underlying the salt passage phenomenon.

Zimbardo (21) subjected this hypothesis to an explicit test. He had students from an introductory psychology class sit at a table near a salt shaker while a confederate would say either "Salt!" or "Assault!" He hypothesized that compliance would be as great in the "Salt!" as in the "Assault!" condition. Zimbardo found, however, that the utterance "Assault!" was met with more calls for clarification than the utterance "Salt!" and the utterance "Assault!" had to be repeated more frequently before the salt would move.[3]

2. However, in a replication of the original Hovland, *et al.* study, Triandis (19) uncovered the opposite tendency due to race. That is, Triandis found that white soldiers were more likely to pass the salt to black soldiers, while black soldiers were more likely to tell the white soldiers to get the salt themselves.

3. In a replication and extension of the Zimbardo experiment, Kelley (11) found that if the confederate had a steak in front of him. "Assault!" was just as effective as "Salt!" in causing salt passage. Kelley concluded that receivers make attributions as to the meaning of utterances based on environmental cues that they perceive.

The search for the source of regularity in salt passing behavior was extended to situational variables.

7   Asch (1) tested the effects of pressure to conform on salt passage. In an experiment, a subject was seated at a table with seven confederates. The subject and six of the confederates had salt shakers in front of them, one confederate did not. The confederate without the salt shaker said, "Please pass the salt." Asch found that, when one confederate passed the salt, the subject was more likely not to pass the salt; but when all the confederates passed the salt, the subject was more likely to conform to peer pressure and also pass the salt. Asch concluded that conformity was an essential aspect of salt passage.

8   Festinger (6) tested the effects of substance uncertainty on salt passage. Subjects were placed at a table where salt was loosely piled on a napkin, while sugar was placed in a salt shaker. When a confederate said, "Please pass the salt," the overwhelming number of subjects passed the sugar. From this study, Festinger concluded that the salt shaker, not the salt itself, was the crucial factor in salt passage.

9   Bem (2) extended Festinger's study by placing two shakers on the table, both clearly marked with the word "SALT." One shaker had salt in it; the other, however, was filled with pepper. Bem reasoned that, if the salt *shaker* were the crucial factor, both the pepper and salt should be passed about an equal number of times. Surprisingly, Bem found that when prompted with the utterance, "Please pass the salt," people more frequently passed the shaker with salt in it than passed the shaker with pepper in it. Bem concluded that, in salt passage, there is an interaction effect between substance in the shaker and the shaker itself.

10   Festinger (7) tested the effects of payment on subject evaluation of salt passage. In a "high reward" condition, subjects were given $20 for passing the salt. In a "low reward" condition, subjects were given $1 for passing the salt. Subjects' evaluations of how much they liked salt passing were then obtained. No significant differences in salt passage liking were found between the two groups. Subjects paid $20, however, expressed more interest in participating in another session of the experiment than did their $1 counterparts. Festinger concluded that subjects in the $1 condition were probably more trustworthy than subjects in the $20 condition.

Milgram (13) tested the effects of threats on salt passage. In a "no 11
threat" condition, subjects were not forewarned about any consequences of passing salt to a confederate. In a "high threat" condition, subjects were told that if they passed the salt, they would be
struck by lightning. Subjects were seated in metal chairs attached
to lightning rods. Thunder in the distance was simulated. Significant differences were found in salt passage compliance between "no
threat" and "high threat" groups. Interestingly, in the "high
threat" group, there was differential response to the threat of
lightning. For golfers and persons who had previously undergone
electroschock therapy, there was less reluctance to exposure to
possible lightning bolts. Milgram concluded that, for most people,
salt passage is contingent on a supportive environment.

*In a descriptive study Schramm (18) reported that the
utterance, "Please pass the salt," was more efficacious
in England, Canada, and the United States, than it was
in Argentina, Pakistan, and Korea.*

Schramm noted the high correlation between the countries 12
where "Please pass the salt" was effective and the degree of exposure of the populace to mass media. He concluded that salt passage
is related to an index of the number of color television sets, tape
cassettes, and Moog synthesizers in a country. Schramm, however,
made no claims about the causal ordering of the variables.

Orne (15) studied the motivations to comply among salt passers. 13
After exposing subjects to the treatments of typical salt passage
studies, he asked them for their motivations in salt passage. Options
were

a. I passed the salt because I thought I would be rewarded.
b. I passed the salt to reduce cognitive dissonance.
c. I passed the salt because the behavior was consistent with
previously made public commitments to salt passing.
d. I passed the salt because that's what I thought I was supposed
to do.

Over 90 percent of all subjects chose response d, strong evidence 14
of the presence of high demand characteristics in the situation.
Responses a, b, and c were more popular among students with social

science backgrounds. Orne cautioned, nonetheless, that the high demand characteristics of these situations may call into question the findings of previous research.

Why does salt move from one end of a table to another when someone says, "Please pass the salt?"

15   Through the efforts of social science researchers, we are able to offer some educated guesses as to the causes of salt passage. Unfortunately, we do not yet have a complete understanding of this complex phenomenon. Findings tend to be inconclusive or inconsistent. Clearly, more research is needed.

16   Future research must be more systematic. Three directions especially warrant pursuit. First, although research to date has uncovered no personality correlates of salt passage compliance, this is probably due to the few numbers of personality traits that have been examined. There are still numerous personality traits left to investigate: Machiavellianism, authoritarianism, social desirability, tendency to embarrass easily, and so on. Possible interaction effects between source and receiver personality characteristics suggest that there are years of necessary research yet to be done in this area.

17   Second, future research needs to be concerned with the effects of demographic variables. The importance of race differences found by Hovland, *et al.* and Triandis cannot be overlooked. (The fact that the Triandis findings conflict with the findings of Hovland, *et al.* should not discourage us, but sensitize us to the complexity of the phenomenon under investigation.) Crucial demographic variables—like sex, age, preferred side of bed for arising in the morning, religion, and others—have yet to be examined.

18   Third, future research needs to be concerned with the effects of situational variables on salt passage. Kelley's "presence of steak" variable and Milgram's "high threat" variable are suggestive. Effects of information-rich environments, overcrowding, presence of armed conflict, and so on would seem to mediate the salt passage phenomenon.

19   Finally, given the complexity of salt passage, social scientists must be willing to abandon their traditional two-variables approach. More sophisticated methodologies are needed. Consideration must be given to using variables from all three research areas

to construct elaborated non-recursive path models permitting both correlated and uncorrelated error terms. Until our methods match the complexity of our phenomena, we are apt to be left with more questions than answers.

In summary, then, we find that at present social science has not found firm evidence to support the validity of the folk belief that the utterance, "Please pass the salt," is causally linked to the movement of salt from one end of a table to another. Salt passage is a complex phenomenon and systematic research on the impact of personality traits, demographics, and situational variables must be assessed. The question of why the utterance, "Please pass the salt," should be associated with salt passage continues to be a source of puzzlement and intrigue for social scientists.

References

1. Asch, R. "Conformity as the Cause of Everything." *Journal of Unique Social Findings* 13, 1952, pp. 62–69.
2. Berm, R. *Beliefs, Attitudes, Values, Mores, Ethics, Existential Concerns, World Views, Notions of Reincarnation and Human Affairs.* Belmont, Cal.: Wadsworth, 1969.
3. Berkeley, R. *It's All in Your Head.* London: Oxford University Press, 1730.
4. Chaucer, R. "The Salt Merchant's Tale." In R. Chaucer (Ed.), *The Canterbury Tales.* London: Cambridge University Press, 1384.
5. Dickens, R. *David Saltmine.* London: Oxford University Press, 1857.
6. Festinger, R. "Let's Take the Salt out of the Salt Shaker and See What Happens." *Journal for Predictions Contrary to Common Sense* 10, 1956, pp. 1–20.
7. Festinger, R. "Let's Give Some Subjects $20 and Some Subjects $1 and See What Happens." *Journal for Predictions Contrary to Common Sense* 18, 1964, pp. 1–20.
8. Hovland, R., R. Lumsdaine, and R. Sheffield, "Praise the Lord and Pass the Salt." *Proceedings of the Academy of Wartime Chaplains* 5, 1949, pp. 13–23.
9. Hume, R. "A Refutation of Berkeley: An Empirical Approach to Salt Passing." *Philosophical Discourse* 278, 1770, pp. 284–296.
10. Janis, R. and R. Feshbach. "Vocal Utterances and Salt Passage: The Importance of the Phonemes in 'Salt'." *Linguistika* 18, 1954, pp. 112–118.
11. Kelley, R. "Attributions Based on Perceived Environmental Cues in Situations of Uncertainty: The Effects of Steak Presence on Salt Passage." *Journal of Pepper and Salt Psychology* 32, 1968, pp. 1–5.

12. McClelland, R. "Brown-nosing and Salt-passing." *Journal for Managerial and Applied Psychology* 18, 1961, pp. 353–362.

13. Milgram, R. "An Electrician's Wiring Guide to Social Science Experiments." *Popular Mechanics* 23, 1969, pp. 74–87.

14. Newcomb, R. "The ABS Model: When S Is Salt." *Journal for Emeritus Ideas* 12, 1958, pp. 10–18.

15. Orne, R. "Salt on Demand: Levels of Moral Reasoning in Salt Passing Behavior." Forthcoming unpublished manuscript.

16. Osgood, R. and R. Tannenbaum. "Taking Requests with a Grain of Salt: Effects of Source Credibility on Salt Passage." *Morton Salt Newsletter* 42, 1953, pp. 2–3.

17. Rokeach, R. *A Whole Earth Catalog of Personality Correlates for the Social Sciences.* New York: It's Academic Press, 1960.

18. Schramm, R. *Process and Effects of Mass Media Extend to Everything.* Frankfurt, Germany: Gutenburg Press, 1970.

19. Triandis, R. "Salt and Pepper: Racial Differences in Salt Passing Behavior." *Journal of Social Findings for Improved Social Relations* 110, 1973, pp. 16–61.

20. Whitehead, R. "A Refutation of Berkeley and Hume: The Need for a Process Perspective of Salt Passage." *Journal of Static Philosophy* 1, 1920, pp. 318–350.

21. Zimbardo, R. "Salt by Any Other Name Is Not Quite So Salty." *Reader's Digest* 38, 1964, pp. 86–114.

STUDY QUESTIONS

1. In paragraph 3, Pencil reports that the presence of other people at the table is a necessary condition for the utterance, "Pass the salt," to be effective. Which of Mill's Methods is used to prove this?

2. In paragraph 5, what method is used to support the hypothesis that the phonemes in "salt" are the causal mechanism? And what method is used in paragraph 6 to undermine that hypothesis?

3. For each of the "experiments" reported in paragraphs 7 through 14, determine what logical methods are being used.

4. The experiments Pencil reports are funny because they involve elaborate efforts to establish something obvious or because they involve testing hypotheses that are wildly off

track. What does this tell us about the process by which
scientists select actual hypotheses to test by experiment?

Reference Chapter: 15

T. G. R. BOWER

The Visual World of Infants

*T. G. R. Bower, born in 1941, received his Ph.D. in psychology
from Cornell University, and is currently Founders Professor
at the School of Human Development, University of Texas at
Dallas. The following selection describes the first of a series of
experiments conducted by Bower to investigate the capacity of
infants to perceive distance. Bower's article follows the classic
model for scientific reporting: The problem is stated, the competing theories are presented, the experimental apparatus is
described, the contrary predictions of the competing theories
are derived, the experimental results are presented, and the implications of the results for the theories are derived.*

What does an infant see as he gazes at the world around him—an   1
ordered array of stable objects or a random flux of evanescent
shadows? There are proponents of both answers. Some psychologists
have maintained that the ability to perceive the world is as much a

[T. G. R. Bower, excerpt from "The Visual World of Infants." *Scientific
American* (December 1966), pp. 80–84.]

part of man's genetic endowment as the ability to breathe; others have contended that perception is an acquired capacity, wholly dependent on experience and learning. The nativists have argued that a baby sees about what adults see; empiricists have held that an infant's visual world must be—in William James's words—"buzzing confusion."

2   At the heart of the argument there is a genuine scientific question: how to account for the discrepancy between the richness of perception and the poverty of its apparent cause—the momentary retinal image. First of all, there is the problem of space perception. The world as perceived seems to have one more dimension—the dimension of depth—than the retinal image does. Then there are the spatial "constancies," the tendency of an object to retain its size regardless of changes in viewing distance (even though the size of the retinal image changes) and to retain its shape even when its orientation (and therefore its retinal image) is changed. In other words, perception seems faithful to the object rather than to its retinal image.

3   Most psychologists have given empiricist answers to the problems of space perception and constancy. They have assumed that the infant's perceptual world mirrors the sequence of momentary retinal images that creates it. The chaotic two-dimensional ensemble of changing shapes is slowly ordered, in this view, by various mechanisms. The retinal image contains many cues to depth; for example, far-off objects are projected lower on the retina than nearby objects (which is why they appear higher to us). Supposedly a baby learns that it must crawl or reach farther to get to such a higher image, and so comes to correlate relative height with relative distance. A similar correlation is presumably made in the case of the many other distance cues.

4   Once these theories have accounted for space perception they must go on to endow infants with the constancies. The oldest theory of constancy learning stems from Hermann von Helmholtz. He argued that by seeing an object at different distances and in a variety of orientations one learns the set of retinal projections that characterize it, so that on encountering a familiar retinal projection one can infer the size, shape, distance and orientation of the object producing it. According to this theory, however, there could be no

constancy with an unfamiliar object. To avoid this prediction a different (but still empiricist) theory was developed from a suggestion made by the Gestalt psychologist Kurt Koffka. It assumes that a child who has acquired space perception will notice that there is a predictable relation between the distance of an image and its size, and a predictable relation between the orientation of an image and its shape. Once these relations have been inferred the child should be able to predict the size an image would have if its distance were changed and the shape it would have if its orientation were changed. The child could achieve shape constancy by predicting what shape a slanted image would have if it were rotated to lie directly across his line of sight. This means that the constancies could be attained with any object, familiar or not.

Note that this theory makes an assertion about the course of perceptual development that is also an assertion about the sequence of events in perception: Before an infant can attain size and shape constancy he must be able to register distance or orientation and projective size or shape; before an adult can compute true size or shape he must register projective size and shape and distance and orientation. There were many attempts to validate this theory of development by testing adult subjects. In a typical experiment adults were shown shapes in various orientations and asked to report true shape, projective shape and orientation. If the theory were correct, one should be able to predict a subject's true-shape judgments from the other two. This turned out to be impossible. Subjects' judgments of true shape were often far more accurate than any deduction from their judgments of projective shape and slant could have been; they often got the true shape right and, say, the orientation completely wrong. This apparent disproof of the theory was explained away by supposing the process of deduction from retinal projection and orientation to true shape had become so automatic with long practice that the premises of the deduction had become subconscious.

So-called completion effects are another puzzle created by the characteristics of the retinal image. If one object is partly occluded by another, the two-dimensional retinal image of the first object is transformed in bizarre ways. Koffka argued that when an adult looks at a book on a table, he can see the table under the book. Similarly, if one looks at a triangle with a pencil across it, one can still see that there is a triangle under the pencil. In the retinal image, of course, there is a gap in the table and a gap in the triangle. Empiricists argue that experience and learning are necessary for these retinal deficiencies to be corrected. They argue that one *learns* the shape of a triangle, after which, on seeing a partly covered triangle, one can infer what the hidden parts look like.

7   This short review of problems will give the reader some idea of what the perceptual world of the infant should be if empiricist theory is correct. William James's description of it as buzzing confusion is then too mild; it should be a chaotic, frightening flux in which nothing stays constant, in which sizes, shapes and edges change, disappear and reappear in a confusing flow. Yet the theories described above are at least serious attempts to handle the problems of space perception, spatial constancies and completion effects. Nativist theories in contrast make a rather poor showing. Too often a nativist theory has merely been an argument against empiricism, not an attempt at genuine explanation.

8   The nativism-empiricism issue has remained open largely because there have seemed to be no ways to investigate the perceptual world of infants. How is one to get an answer from an organism as helpless as a young human infant, capable of few responses of any kind and of even fewer spatially directed responses? One obvious solution is the use of the operant-conditioning methods that were devised by B. F. Skinner: one selects some response from an organism's repertory and delivers some "reinforcing" agent contingent on the occurrence of that response. If reinforcement is delivered only in the presence of a certain stimulus, the response soon occurs only in the presence of that "conditioned" stimulus. It is then possible, among other things, to introduce new stimuli to be discriminated from the conditioned stimulus. Using manipulations of this kind, one can discover a great deal about the perceptual worlds of pigeons, fishes and even worms.

9   For some years, first at Cornell University and more recently at Harvard University, I have been applying operant-conditioning techniques to investigate perception in human infants. The major block to applying these methods to infants had been the necessity of a reinforcing agent. The agent is ordinarily either food or water, and it is usually withheld for some time before the experiment; a

pigeon working for grain is kept at 80 percent of its normal body weight. One could hardly inflict such privation on infants. Fortunately less drastic methods are available. As Skinner pointed out, any change in surroundings, even the simple rustling of a newspaper, seems to reinforce an infant's responses. The reinforcement we use is a little game that adults often play with infants called "peekaboo": the adult pops out in front of the infant, smiling and nodding, and speaks to him (sometimes patting him on the tummy if he is unresponsive) and then quickly disappears from view. Infants between two and 20 weeks old seem to find this event highly reinforcing and will respond for 20 minutes at a time to make it occur. The situation can be made even more reinforcing by altering the schedule of reinforcement.

A lesser problem is deciding what response to use. Infants have few responses available; most of these require substantial effort and would quickly tire the subject of an experiment. The one used in the present investigations is a turn of the head. The infant reclines with his head between two yielding pads. By turning his head as little as half an inch to the left or right he closes a microswitch that operates a recorder. This response requires scant effort; even infants as young as two weeks old can give 400 such responses with no apparent fatigue. ₁₀

The first experiment carried out with these techniques was aimed at discovering whether or not infants can perceive distance and are capable of size constancy. An infant between six and eight weeks old, too young to be capable of the spatial behavior of reaching and crawling, reclined in an infant seat on a table, with the peekabooing experimenter crouching in front of him and the stimuli beyond the experimenter. A translucent screen could be raised for rest periods and stimulus changes. The conditioned stimulus in the first experiment was a white cube 30 centimeters (12 inches) on a side, placed one meter from the infant's eyes. ₁₁

After training an infant to respond only in the presence of the cube, we gradually changed the reinforcement schedule to a variable one in which every fifth response, on the average, was reinforced. After one experimental hour on this schedule we began perceptual testing by introducing three new stimuli. These were the 30-centimeter cube placed three meters away, a 90-centimeter cube ₁₂

placed one meter away and the same 90-centimeter cube placed three meters away. These three stimuli and the conditioned stimulus were each presented for four 30-second periods in counterbalanced order and the number of responses elicited by each stimulus was recorded. During the testing period no reinforcement was given.

13   On any theory, the conditioned stimulus could be expected to elicit more responses than any of the other stimuli. The stimulus eliciting the next highest number of responses should be the one that appears to the infant to be most like the conditioned stimulus. If the empiricist hypothesis that infants do not perceive distance and do not have size constancy is correct, the stimulus that should have appeared most similar was the third stimulus, the 90-centimeter cube placed three meters away; it was three times the height and width of the conditioned stimulus but also three times as far away, so that it projected a retinal image of the same size. If infants can perceive distance but still lack size constancy, stimulus 3 should still have seemed more like the conditioned stimulus than stimulus 1, the 30-centimeter cube at three meters. Both were at the same distance, but stimulus 3 projected a retinal image with the same area as the conditioned stimulus, whereas stimulus 1 projected an image with only one-ninth the area. If the infants had been unable to discriminate distance at all, stimulus 2 would have elicited as many responses as stimulus 1, since stimulus 2 projected an image with nine times the area of the conditioned stimulus and stimulus 1 projected an image one-ninth as large. If they had been sensitive to distance or its cues but had lacked size constancy, stimulus 2 would have elicited more responses than stimulus 1, since stimulus 2 was at the same distance as the conditioned stimulus. If, on the other hand, the infants had been able to perceive distance and had size constancy, stimuli 1 and 2 should have elicited about the same number of responses, since stimulus 1 differed from the conditioned stimulus in distance and stimulus 2 differed in size; stimulus 3 should have elicited the lowest number of responses, since it differed from the conditioned stimulus in both size and distance [see figures].

14   To sum up the predictions, according to empiricism stimulus 3 should be as effective as or more effective than stimulus 2, which should in turn get more responses than stimulus 1; if retinal distance cues are not taken into account, 3 should be clearly superior to 2.

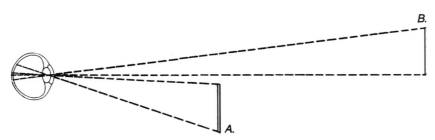

Retinal Image varies with distance. Objects *A* and *B* are the same size but *A* projects a larger image on the retina. In spite of this, adults usually see such objects as being the same size. One of the author's objectives was to see if infants too have this "size constancy."

CONDITIONED STIMULUS	TEST STIMULI 1	2	3
TRUE SIZE			
TRUE DISTANCE 1	3	1	3
RETINAL SIZE			
RETINAL DISTANCE CUES	DIFFERENT	SAME	DIFFERENT

Size Constancy was investigated with cubes of different sizes placed at different distances from the infants. The conditioned stimulus was 30 centimeters on a side and one meter away, test stimuli 30 or 90 centimeters on a side and one or three meters away. The chart shows how test stimuli were related to the conditioned stimulus in various respects.

According to nativism the order should be the reverse: more responses to 1 than to 2, more to 2 than to 3. What happened was that the conditioned stimulus elicited an average of 98 responses; stimulus 1, 58 responses; stimulus 2, 54 responses, and stimulus 3, 22 responses. It therefore seems that the retinal image theory cannot be correct. These infants' responses were affected by real size and real distance, not by retinal size or by retinal distance cues.

STUDY QUESTIONS

1. The experimental apparatus allows Bower to manipulate three variables: size of object, distance of object from infant, and thereby the size of the retinal image. What is the distinction Bower makes between the perception of distance and size constancy? How does the experimental apparatus allow him to test whether they are within the infant's capabilities?
2. What is the empiricist theory that Bower wants to test, and how does it differ from the nativist theory?
3. In paragraph 13, Bower derives five predictions. Which of the predictions are predictions of the empiricist theory, and which of the nativist theory?
4. What is the retinal image theory Bower rejects in the final paragraph, and why does he think the experimental results allow him to reject it?
5. Which of Mill's Methods is Bower following?

Reference Chapter: 15

ALEXANDER PETRUNKEVITCH

The Spider and the Wasp

Alexander Petrunkevitch (1875–1964) was born in Russia, and was Professor of Zoology at Yale University. He was the author of several books, including Principles of Classification *and* Index Catalog of Spiders of North, Central, and South America. *In the following essay, published originally in* Scientific American, *Petrunkevitch discusses a puzzling relationship between tarantulas and the digger wasp Pepsis.*

In the feeding and safeguarding of their progeny insects and spiders exhibit some interesting analogies to reasoning and some crass examples of blind instinct. The case I propose to describe here is that of the tarantula spiders and their archenemy, the digger wasps of the genus Pepsis. It is a classic example of what looks like intelligence pitted against instinct—a strange situation in which the victim, though fully able to defend itself, submits unwittingly to its destruction.

Most tarantulas live in the tropics, but several species occur in the temperate zone and a few are common in the southern U.S. Some varieties are large and have powerful fangs with which they can inflict a deep wound. These formidable looking spiders do not, however, attack man; you can hold one in your hand, if you are gentle, without being bitten. Their bite is dangerous only to insects

[Alexander Petrunkevitch, "The Spider and the Wasp." *Scientific American* (August 1952), pp. 21–23.]

and small mammals such as mice; for man it is no worse than a hornet's sting.

3   Tarantulas customarily live in deep cylindrical burrows, from which they emerge at dusk and into which they retire at dawn. Mature males wander about after dark in search of females and occasionally stray into houses. After mating, the male dies in a few weeks, but a female lives much longer and can mate several years in succession. In a Paris museum is a tropical specimen which is said to have been living in captivity for 25 years.

4   A fertilized female tarantula lays from 200 to 400 eggs at a time; thus it is possible for a single tarantula to produce several thousand young. She takes no care of them beyond weaving a cocoon of silk to enclose the eggs. After they hatch, the young walk away, find convenient places in which to dig their burrows and spend the rest of their lives in solitude. The eyesight of tarantulas is poor, being limited to a sensing of change in the intensity of light and to the perception of moving objects. They apparently have little or no sense of hearing, for a hungry tarantula will pay no attention to a loudly chirping cricket placed in its cage unless the insect happens to touch one of its legs.

5   But all spiders, and especially hairy ones, have an extremely delicate sense of touch. Laboratory experiments prove that tarantulas can distinguish three types of touch: pressure against the body wall, stroking of the body hair, and riffling of certain very fine hairs on the legs called trichobothria. Pressure against the body, by the finger or the end of a pencil, causes the tarantula to move off slowly for a short distance. The touch excites no defensive response unless the approach is from above where the spider can see the motion, in which case it rises on its hind legs, lifts its front legs, opens its fangs and holds this threatening posture as long as the object continues to move.

6   The entire body of a tarantula, especially its legs, is thickly clothed with hair. Some of it is short and wooly, some long and stiff. Touching this body hair produces one of two distinct reactions. When the spider is hungry, it responds with an immediate and swift attack. At the touch of a cricket's antennae the tarantula seizes the insect so swiftly that a motion picture taken at the rate of 64 frames per second shows only the result and not the process of capture. But when the spider is not hungry, the stimulation of its hairs merely

causes it to shake the touched limb. An insect can walk under its hairy belly unharmed.

The trichobothria, very fine hairs growing from dislike[1] membranes on the legs, are sensitive only to air movement. A light breeze makes them vibrate slowly, without disturbing the common hair. When one blows gently on the trichobothria, the tarantula reacts with a quick jerk of its four front legs. If the front and hind legs are stimulated at the same time, the spider makes a sudden jump. This reaction is quite independent of the state of its appetite.

7

These three tactile responses—to pressure on the body wall, to moving of the common hair, and to flexing of the trichobothria—are so different from one another that there is no possibility of confusing them. They serve the tarantula adequately for most of its needs and enable it to avoid most annoyances and dangers. But they fail the spider completely when it meets its deadly enemy, the digger wasp Pepsis.

8

These solitary wasps are beautiful and formidable creatures. Most species are either a deep shiny blue all over, or deep blue with rusty wings. The largest have a wing span of about four inches. They live on nectar. When excited, they give off a pungent odor—a warning that they are ready to attack. The sting is much worse than that of a bee or common wasp, and the pain and swelling last longer. In the adult stage the wasp lives only a few months. The female produces but a few eggs, one at a time at intervals of two or three days. For each egg the mother must provide one adult tarantula, alive but paralyzed. The mother wasp attaches the egg to the paralyzed spider's abdomen. Upon hatching from the egg, the larva is many hundreds of times smaller than its living but helpless victim. It eats no other food and drinks no water. By the time it has finished its single Gargantuan meal and become ready for wasphood, nothing remains of the tarantula but its indigestible chitinous skeleton.

9

The mother wasp goes tarantula-hunting when the egg in her ovary is almost ready to be laid. Flying low over the ground late on a sunny afternoon, the wasp looks for its victim or for the mouth of a tarantula burrow, a round hole edged by a bit of silk. The sex of the spider makes no difference, but the mother is highly discrimi-

10

1. Unlike or dissimilar.

nating as to species. Each species of Pepsis requires a certain species of tarantula, and the wasp will not attack the wrong species. In a cage with a tarantula which is not its normal prey, the wasp avoids the spider and is usually killed by it in the night.

11   Yet when a wasp finds the correct species, it is the other way about. To identify the species the wasp apparently must explore the spider with her antennae. The tarantula shows an amazing tolerance to this exploration. The wasp crawls under it and walks over it without evoking any hostile response. The molestation is so great and so persistent that the tarantula often rises on all eight legs, as if it were on stilts. It may stand this way for several minutes. Meanwhile the wasp, having satisfied itself that the victim is of the right species, moves off a few inches to dig the spider's grave. Working vigorously with legs and jaws, it excavates a hole 8 to 10 inches deep with a diameter slightly larger than the spider's girth. Now and again the wasp pops out of the hole to make sure that the spider is still there.

12   When the grave is finished, the wasp returns to the tarantula to complete her ghastly enterprise. First she feels it all over once more with her antennae. Then her behavior becomes more aggressive. She bends her abdomen, protruding her sting, and searches for the soft membrane at the point where the spider's legs join its body— the only spot where she can penetrate the horny skeleton. From time to time, as the exasperated spider slowly shifts ground, the wasp turns on her back and slides along with the aid of her wings, trying to get under the tarantula for a shot at the vital spot. During all this maneuvering, which can last for several minutes, the tarantula makes no move to save itself. Finally the wasp corners it against some obstruction and grasps one of its legs in her powerful jaws. Now at last the harassed spider tries a desperate but vain defense. The two contestants roll over and over on the ground. It is a terrifying sight and the outcome is always the same. The wasp finally manages to thrust her sting into the soft spot and holds it there for a few seconds while she pumps in the poison. Almost immediately the tarantula falls paralyzed on its back. Its legs stop twitching; its heart stops beating. Yet it is not dead, as is shown by the fact that if taken from the wasp it can be restored to some sensitivity by being kept in a moist chamber for several months.

13   After paralyzing the tarantula, the wasp cleans herself by drag-

ging her body along the ground and rubbing her feet, sucks the drop
of blood oozing from the wound in the spider's abdomen, then grabs
a leg of the flabby, helpless animal in her jaws and drags it down to
the bottom of the grave. She stays there for many minutes, sometimes for several hours, and what she does all that time in the dark
we do not know. Eventually she lays her egg and attaches it to the
side of the spider's abdomen with a sticky secretion. Then she
emerges, fills the grave with soil carried bit by bit in her jaws, and
finally tramples the ground all around to hide any trace of the grave
from prowlers. Then she flies away, leaving her descendant safely
started in life.

In all this the behavior of the wasp evidently is qualitatively   14
different from that of the spider. The wasp acts like an intelligent
animal. This is not to say that instinct plays no part or that she
reasons as man does. But her actions are to the point; they are not
automatic and can be modified to fit the situation. We do not know
for certain how she identifies the tarantula—probably it is by some
olfactory or chemo-tactile sense—but she does it purposefully and
does not blindly tackle a wrong species.

On the other hand, the tarantula's behavior shows only confu-   15
sion. Evidently the wasp's pawing gives it no pleasure, for it tries to
move away. That the wasp is not simulating sexual stimulation is
certain because male and female tarantulas react in the same way
to its advances. That the spider is not anesthetized by some odorless
secretion is easily shown by blowing lightly at the tarantula and
making it jump suddenly. What, then, makes the tarantula behave
as stupidly as it does?

No clear, simple answer is available. Possibly the stimulation by   16
the wasp's antennae is masked by a heavier pressure on the spider's
body, so that it reacts as when prodded by a pencil. But the explanation may be much more complex. Initiative in attack is not in the
nature of tarantulas; most species fight only when cornered so that
escape is impossible. Their inherited patterns of behavior apparently prompt them to avoid problems rather than attack them. For
example, spiders always weave their webs in three dimensions, and
when a spider finds that there is insufficient space to attach certain
threads in the third dimension, it leaves the place and seeks another,
instead of finishing the web in a single plane. This urge to escape
seems to arise under all circumstances, in all phases of life, and to

take the place of reasoning. For a spider to change the pattern of its web is as impossible as for an inexperienced man to build a bridge across a chasm obstructing his way.

17   In a way the instinctive urge to escape is not only easier but often more efficient than reasoning. The tarantula does exactly what is most efficient in all cases except in an encounter with a ruthless and determined attacker dependent for the existence of her own species on killing as many tarantulas as she can lay eggs. Perhaps in this case the spider follows its usual pattern of trying to escape, instead of seizing and killing the wasp, because it is not aware of its danger. In any case, the survival of the tarantula species as a whole is protected by the fact that the spider is much more fertile than the wasp.

STUDY QUESTIONS

1. Paragraphs 2 to 8 offer a lengthy description of the tarantula's lifestyle and capacities. What is Petrunkevitch's purpose in presenting such a lengthy description?
2. What features of the interaction between the spider and the wasp are the phenomena to be explained?
3. What provisional explanation does Petrunkevitch accept for the tarantula's behaving as "stupidly" as it does?
4. How many other possible explanations for the tarantula's behavior does Petrunkevitch consider? And on what grounds does he reject those other explanations?
5. In the first sentence of the essay, Petrunkevitch suggests that the actions of "insects and spiders exhibit some interesting analogies to reasoning," and in paragraph 14 he suggests that the wasp "acts like an intelligent animal." What seems to be intelligent about the wasp's actions? How would one determine whether the wasp's actions really are intelligent actions or only analogous to intelligent actions?

Reference Chapter: 18

PART SIX

On Education

JEAN-JACQUES ROUSSEAU

Children Should Not Be Reasoned With

Jean-Jacques Rousseau (1712–1778) was a Swiss-born French intellectual. Through two widely read books, Emile, *from which the following passage is excerpted, and* The Social Contract *(both published 1762), Rousseau became an influential forerunner of Romanticism, a nineteenth-century intellectual and artistic movement. One of the major themes of Romanticism, which the following passage illustrates, is that feeling and instinct are much more important than reason as guides to human action.*

[J.-J. Rousseau, excerpt from *Emile*, transl. Eleanor Worthington. Boston: Ginn, Heath & Co., 1883, pp. 52–54.]

1   *Reasoning should not begin too soon*—Locke's[1] great maxim was that we ought to reason with children, and just now this maxim is much in fashion. I think, however, that its success does not warrant its reputation, and I find nothing more stupid than children who have been so much reasoned with. Reason, apparently a compound of all other faculties, the one latest developed, and with most difficulty, is the one proposed as agent in unfolding the faculties earliest used! The noblest work of education is to make a reasoning man, and we expect to train a young child by making him reason! This is beginning at the end; this is making an instrument of a result. If children understood how to reason they would not need to be educated. But by addressing them from their tenderest years in a language they cannot understand, you accustom them to be satisfied with words, to find fault with whatever is said to them, to think themselves as wise as their teachers, to wrangle and rebel. And what we mean they shall do from reasonable motives we are forced to obtain from them by adding the motive of avarice, or of fear, or of vanity.

2   Nature intends that children shall be children before they are men. If we insist on reversing this order we shall have fruit early indeed, but unripe and tasteless, and liable to early decay; we shall have young savants and old children. Childhood has its own methods of seeing, thinking, and feeling. Nothing shows less sense than to try to substitute our own methods for these. I would rather require a child ten years old to be five feet tall than to be judicious. Indeed, what use would he have at that age for the power to reason? It is a check upon physical strength, and the child needs none.

3   In attempting to persuade your pupils to obedience you add to this alleged persuasion force and threats, or worse still, flattery and promises. Bought over in this way by interest, or constrained by force, they pretend to be convinced by reason. They see plainly that as soon as you discover obedience or disobedience in their conduct, the former is an advantage and the latter a disadvantage to them. But you ask of them only what is distasteful to them; it is always irksome to carry out the wishes of another, so by stealth they carry out their own. They are sure that if their disobedience is not known they are doing well; but they are ready, for fear of greater evils, to

---

1. John Locke (1632–1704), English philosopher.—Eds.

acknowledge, if found out, that they are doing wrong. As the reason for the duty required is beyond their capacity, no one can make them really understand it. But the fear of punishment, the hope of forgiveness, your importunity, their difficulty in answering you, extort from them the confession required of them. You think you have convinced them, when you have only wearied them out or intimidated them.

What results from this? First of all that, by imposing upon them a duty they do not feel as such, you set them against your tyranny, and dissuade them from loving you; you teach them to be dissemblers, deceitful, wilfully untrue, for the sake of extorting rewards or of escaping punishments. Finally, by habituating them to cover a secret motive by an apparent motive, you give them the means of constantly misleading you, of concealing their true character from you, and of satisfying yourself and others with empty words when their occasion demands. You may say that the law, although binding on the conscience, uses constraint in dealing with grown men. I grant it; but what are these men but children spoiled by their education? This is precisely what ought to be prevented. With children use force, with men reason; such is the natural order of things. The wise man requires no laws.

4

STUDY QUESTIONS

1. Diagram the overall structure of Rousseau's argument.
2. What does Rousseau seem to mean by "reason" and "reasoning"?
3. In paragraphs 3 and 4, Rousseau argues that in attempting to teach children to reason, one ends up teaching them hypocrisy. What capacities does he assume children have that makes them capable of being hypocritical?
4. How many other bad consequences does Rousseau attribute to teaching children to reason?
5. What alternative method of dealing with children does Rousseau propose? Is this the only alternative?

Reference Chapters: 7, 11

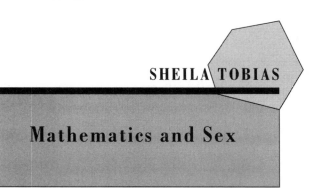

SHEILA TOBIAS

Mathematics and Sex

Sheila Tobias was Associate Provost at Wesleyan University. In this selection, taken from her book Overcoming Math Anxiety, *she evaluates various explanations for the disparity between boys' and girls' math skills.*

1   Men are not free to avoid math; women are.

2   In a major address to the American Academy of Arts and Sciences in 1976, Gerard Piel, publisher of *Scientific American,* cited some of the indicators of mathematics avoidance among girls and young women. "The SAT record plainly suggests that men begin to be separated from women in high school," he noted. "At Andover [an elite private high school] 60 percent of the boys take extra courses in both mathematics and science, but only 25 percent of the girls. . . . By the time the presently graduating high school classes are applying to graduate school," he concluded, "only a tenth as many young women as men will have retained the confidence and capacity to apply to graduate study in the sciences.

3   Some other measures of mathematics avoidance among females are these:

> Girls accounts for 49 percent of the secondary school students in the United States but comprise only 20 percent of those taking math beyond geometry.

[Sheila Tobias, excerpts from Chapter 3 of *Overcoming Math Anxiety.* New York: W. W. Norton, 1978, pp. 70–72, 74, 77–88, 91–96, some footnotes, illustrations, and cross-references deleted; second edition available 1994.]

The college and university population totals 45 percent women, yet only 15 percent of the majors in pure mathematics are women.
Women make up 47 percent of the labor force and 42 percent of those engaged in professional occupations. Yet they are only 12 percent of the scientific and technical personnel working in America today.

Are these data simply evidence of individual preference, or do they represent a pattern of math avoidance and even math anxiety among women? We know that there are differences in *interest* between the sexes. What we do not know is what causes such differences, that is whether these are differences in ability, differences in attitude, or both; and, even more important, whether such differences, if indeed they exist, are innate or learned.

Most learning psychologists doing research today are environmentalists; that is, they tend to be on the "nurture" side of the nature-nurture controversy. Most of them would therefore not subscribe to the man on the street's belief that mathematics ability is just one of those innate differences between men and women that can neither be ignored nor explained away. Yet even the most recent research on sex differences in intelligence accepts the fact that performance in math varies by gender. Because this is assumed to be natural and inevitable (if not genetic in origin) for a long time the causes of female underachievement in mathematics have not been considered a promising area for study and certainly not an urgent one.

But recently, as women began to aspire to positions in fields previously dominated by men, this attitude began to change. The women's movement and the accompanying feminist critique of social psychology can be credited, I believe, with the rise in interest in mathematics and sex and with the formulation of some important new questions. Do girls do poorly in math because they are afraid that people (especially boys) will think them abnormal if they do well, or is it because girls are not taught to believe that they will ever need mathematics? Are there certain kinds of math that girls do better? Which kinds? At what ages? Are there different ways to explain key concepts of math that would help some girls understand them better? . . .

In fact, math avoidance is not just a female phenomenon. Most people of both sexes stop taking math before their formal education is complete. Few people become mathematicians and many very smart people do not like math at all. Thus, "dropping out" of math is nearly universal, and is by no means restricted to girls and women. From this perspective, girls who avoid math and math-related subjects may simply be getting the message sooner than boys that math is unrewarding and irrelevant, but boys will also get that message in time.

8   A recent survey of attitudes toward math among ninth and twelfth graders demonstrated this point very well. Although ninth grade girls had a more negative attitude toward math than ninth grade boys, by the twelfth grade boys had caught up. The researcher concluded that by age 17 a majority of all students have developed an aversion to math, which is tragic but certainly not sex-related.

9   What then is gender-related? What can we say with certainty about mathematics and sex? . . .

10   Popular wisdom holds that females are better at computation and males at problem solving, females at "simple repetitive tasks" and males at restructuring complex ideas. However, since experts cannot even agree on what these categories are, still less on how to measure them, we have to be careful about accepting sex differences in "mathematical reasoning" or "analytic ability" as reported by the researchers in this field. It is fascinating to speculate that there are "innate capacities" to analyze or to reason mathematically, but these qualities have simply not been found.

11   What then do we know? As of 1978, are there any "facts" about male-female differences in mathematics performance that we can accept from the varied and not always consistent research findings? Possibly not, since the field is so very much in flux. But at least until recently, the "facts" were taken to be these:

12   Boys and girls may be born alike in math ability, but certain sex differences in performance emerge as early as such evidence can be gathered and remain through adulthood. They are:

1. Girls compute better than boys (elementary school and on).
2. Boys solve word problems better than girls (from age 13 on).
3. Boys take more math than girls (from age 16 on).
4. Girls learn to hate math sooner and possibly for different reasons.

13   One reason for the differences in performance, to be explored later in this chapter, is the amount of math learned and used at play.

Another may be the difference in male-female maturation.[1] If girls do better than boys at all elementary school tasks, then they may compute better only because arithmetic is part of the elementary school curriculum. As boys and girls grow older, girls are under pressure to become less competitive academically. Thus, the falling off of girls' math performance from age 10 to 15 may be the result of this kind of scenario:

1. Each year math gets harder and requires more work and commitment.
2. Both boys and girls are pressured, beginning at age 10, not to excel in areas designated by society as outside their sex-role domain.
3. Girls now have a good excuse to avoid the painful struggle with math; boys don't.

Such a model may explain girls' lower achievement in math overall, but why should girls have difficulty in problem solving? In her 1964 review of the research on sex difference, Eleanor Maccoby also noted that girls are generally more conforming, more suggestible, and more dependent upon the opinion of others than boys (all learned, not innate behaviors).[2] Thus they may not be as willing to take risks or to think for themselves, two necessary behaviors for solving problems. Indeed, a test of third graders that cannot yet be cited found girls nowhere near as willing to estimate, to make judgments about "possible right answers," and to work with systems they had never seen before. Their very success at doing the expected seems very much to interfere with their doing something new.

If readiness to do word problems, to take one example, is as much a function of readiness to take risks as it is of "reasoning ability," then there is more to mathematics performance than memory, computation, and reasoning. The differences between boys and girls—no matter how consistently they show up—cannot simply be attributed to differences in innate ability.

Still, if you were to ask the victims themselves, people who have

1. Girls are about two years ahead of boys on most indices of biological maturation throughout childhood.

2. This is confirmed by Susan Auslander of the Wesleyan Math Clinic, whose "Analysis of Changing Attitudes toward Mathematics" (1978, unpublished) found that females place more value on outside opinion of success in mathematics than males.

trouble doing math, they would probably not agree; they would say that it has to do with the way they are "wired." They feel that they somehow lack something—one ability or several—that other people have. Although women want to believe they are not mentally inferior to men, many fear that in math they really are. Thus, we must consider seriously whether there is any biological basis for mathematical ability, not only because some researchers believe there is, but because some victims agree with them.

The Arguments from Biology

17 The search for some biological basis for math ability or disability is fraught with logical and experimental difficulties. Since not all math underachievers are women and not all women avoid mathematics, it is not very likely on the face of it that poor performance in math can result from some genetic or hormonal difference between the sexes. Moreover, no amount of speculation so far has unearthed a "mathematical competency" in some tangible, measurable substance in the body. Since masculinity cannot be injected into women to see whether it improves their mathematics, the theories that attribute such ability to genes or hormones must depend on circumstantial evidence for their proof. To explain the percent of Ph.D.'s in mathematics earned by women, we would have to conclude either that these women have different genes, hormones, and brain organization than the rest of us; or that certain positive experiences in their lives have largely undone the negative influence of being female; or both. . . .

18 At the root of many of the assumptions about biology and intelligence is the undeniable fact that there have been fewer women "geniuses." The distribution of genius, however, is more a social than a biological phenomenon. An interesting aspect of the lives of geniuses is precisely their dependence on familial, social, and institutional supports. Without schools to accept them, men of wealth to commission their work, colleagues to talk to, and wives to do their domestic chores, they might have gone unrecognized—they might not even have been so smart. In a classic essay explaining why we have so few great women artists, Linda Nochlin Pommer tells us

that women were not allowed to attend classes in art schools because of the presence of nude (female) models. Nor were they given apprenticeships or mentors; and even when they could put together the materials they needed to paint or sculpt, they were not allowed to exhibit their work in galleries or museums.

Women in mathematics fared little better. Emmy Noether, who may be the only woman mathematician considered a genius, was honored (or perhaps mocked) during her lifetime by being called "Der Noether" ("Der" being the masculine form of "the"). Der Noether notwithstanding, the search for the genetic and hormonal origins of math ability goes on.   19

Genetically, the only difference between males and females (albeit a significant and pervasive one) is the presence of two chromosomes designated "X" in every female cell. Normal males have an "X-Y" combination. Since some kinds of mental retardation are associated with sex-chromosomal anomalies, a number of researchers have sought a link between specific abilities and the presence or absence of the second "X." But the link between genetics and mathematics is simply not supported by conclusive evidence.   20

Since intensified hormonal activity begins at adolescence and since, as we have noted, girls seem to lose interest in mathematics during adolescence, much more has been made of the unequal amounts of the sex-linked hormones, androgen and estrogen, in females and males. Estrogen is linked with "simple repetitive tasks" and androgen, with "complex restructuring tasks." The argument here is not only that such specific talents are biologically based (probably undemonstrable) but also that such talents are either-or; that one cannot be good at *both* repetitive and restructuring kinds of assignments.   21

Further, if the sex hormones were in any way responsible for our intellectual functioning, we should get dumber as we get older since our production of both kinds of sex hormones decreases with age.[3] But as far as we know, hormone production responds to mood,   22

---

3. Indeed, some people do claim that little original work is done by mathematicians once they reach age 30. But a counter explanation is that creative work is done not because of youth but because of "newness to the field." Mathematicians who originate ideas at 25, 20, and even 18 are benefiting not so much from hormonal vigor as from freshness of viewpoint and willingness to ask new questions. I am indebted to Stuart Gilmore, historian of science, for this idea.

activity level, and a number of other external and environmental conditions as well as to age. Thus, even if one day we were to find a sure correlation between the amount of hormone present and the degree of mathematical competence, we would not know whether it was the mathematical competence that caused the hormone level to increase or the hormone level that gave us the mathematical competence.

23   All this criticism of the biological arguments does not imply that what women do with their bodies has no effect on their mathematical skills. As we will see, toys, games, sports, training in certain cognitive areas, and exercise and experience may be the intervening variables we have previously mistaken for biological cause. But first we must look a little more closely at attitude.

Sex Roles and Mathematics Competence

24   The frequency with which girls tend to lose interest in math just at puberty (junior high school) suggests that puberty might in some sense cause girls to fall behind in math. Several explanations come to mind: the influence of hormones, more intensified sex-role socialization, or some extracurricular learning experience boys have at that age that girls do not have. Having set aside the argument that hormones operate by themselves, let us consider the other issues. Here we enter the world of attitudes, as formed by experience and expectation.

25   One group of seventh graders in a private school in New England gave a clue to what children themselves think about this. When visitors to their math class asked why girls do as well as boys in math until sixth grade but after sixth grade boys do better, the girls responded: "Oh, that's easy. After sixth grade, we have to do real math." The reason why "real math" should be considered accessible to boys and not to girls cannot be found in biology, but only in the ideology of sex differences.

26   Parents, peers, and teachers forgive a girl when she does badly in math at school, encouraging her to do well in other subjects instead.

"'There, there,' my mother used to say when I failed at math," one woman remembers. "But I got a talking-to when I did badly in French." "Mother couldn't figure out a 15 percent tip and Daddy seemed to love her more for her incompetence," remembers another. Lynn Fox, who has worked intensively in a program for mathematically gifted teenagers who are brought to the campus of Johns Hopkins University for special instruction, finds it difficult to recruit girls and to keep them in her program. Their parents sometimes prevent them from participating altogether for fear it will make their daughters too different, and the girls themselves often find it difficult to continue with mathematics, she reports, because they experience social ostracism. The math anxious girl we met in Chapter Two, who would have lost her social life if she had asked an interesting question in math class, was anticipating just that.

Where do these attitudes come from? 27

A study of the images of males and females in children's text- 28 books by sociologist Lenore Weitzman of the University of California at Davis, provides one clue to why math is associated with men and boys in the minds of little children. "Two out of every three pictures in the math books surveyed were of males, and the examples given of females doing math were insulting and designed to reinforce the worst of the stereotypes," she reports.

Weitzman comments: "It seems ironic that housewives who use 29 so much math in balancing their accounts and in managing household budgets are shown as baffled by simple addition."

"Another feature of the mathematics textbooks," says 30 Weitzman, "is the frequent use of sex as a category for dividing people, especially for explaining set theory."

"When sex is used as a category, girls are told that they can be 31 classified as different," Weitzman believes, "as typically emotional or domestic . . . There is also strong sex typing in the examples used and in the math problems."

"We found math problems," Weitzman writes, "in which girls 32 were paid less than boys for the same work. It would be hard to imagine a textbook publisher allowing this example if a black boy were being paid less than a white boy. Yet it seems legitimate to underpay girls."

In another survey of math textbooks published in 1969, not one 33 picture of a girl was found and the arithmetic problems used as examples in the book showed adult women having to ask even their children for help with math, or avoiding the task entirely by saying, "Wait until your father comes home."

34   Adults remember their junior high school experiences in math as full of clues that math was a male domain. No so long ago, one junior high school regional math competition offered a tie clasp for first prize. A math teacher in another school, commenting unfavorably on the performance at the blackboard of a male student, said to him, "You think like a girl." If poor math thinkers think like girls, who are good math thinkers supposed to be? . . .

Street Mathematics: Things, Motion, and Scores

35   If a ballplayer is batting .233 going into a game and gets three hits in four times at bat (which means he has batted .750 for the day), someone watching the game might assume that the day's performance will make a terrific improvement in his batting average. But it turns out that the three-for-four day only raises the .233 to .252. Disappointing, but a very good personal lesson in fractions, ratios, and percents.

36   Scores, performances like this one, lengths, speeds of sprints or downhill slaloms are expressed in numbers, in ratios, and in other comparisons. The attention given to such matters surely contributes to a boy's familiarity with simple arithmetic functions, and must convince him, at least on some subliminal level, of the utility of mathematics. This does not imply that every boy who handles runs batted in and batting averages well during the game on Sunday will see the application of these procedures to his Monday morning school assignment. But handling figures as people do in sports probably lays the groundwork for using figures later on.

37   Not all the skills necessary for mathematics are learned in school. Measuring, computing, and manipulating objects that have dimensions and dynamic properties of their own are part of everyday life for some children. Other children who miss these experiences may not be well primed for math in school.

Feminists have complained for a long time that playing with   38
dolls is one way to convince impressionable little girls that they may
only be mothers or housewives, or, in emulation of the Barbie doll,
pinup girls when they grow up. But doll playing may have even
more serious consequences. Have you ever watched a little girl play
with a doll? Most of the time she is talking and not doing, and even
when she is doing (dressing, undressing, packing the doll away) she
is not learning very much about the world. Imagine her taking a
Barbie doll apart to study its talking mechanism. That's not the sort
of thing she is encouraged to do. Do girls find out about gravity and
distance and shapes and sizes playing with dolls? Probably not!

A college text written for inadequately prepared science students   39
begins with a series of supposedly simple problems dealing with
marbles, cylinders, poles made of different substances, levels, balances, and an inclined plane. Even the least talented male science
student will probably be able to see these items as objects, each
having a particular shape, size, and style of movement. He has
balanced himself or some other object on a teeter-totter; he has
watched marbles spin and even fly. He has probably tried to fit one
pole of a certain diameter inside another, or used a stick to pull up
another stick, learning leverage. Those trucks little boys clamor for
and get are moving objects. Things in little boys' lives drop and spin
and collide and even explode sometimes.

The more curious boy will have taken apart a number of house-   40
hold and play objects by the time he is ten; if his parents are lucky,
he may even have put them back together again. In all this he is
learning things that will be useful in physics and math. Taking out
parts that have to go back in requires some examination of form.
Building something that stays up or at least stays put for some time
involves working with structure. Perhaps the absence of things that
move in little girls' childhoods (especially if they are urban little
girls) quite as much as the presence of dolls makes the quantities
and relationships of math so alien to them.

In sports played, as well as sports watched, boys learn more   41
math-related concepts than girls do. Getting to first base on a not
very well hit grounder is a lesson in time, speed, and distance.
Intercepting a football in the air requires some rapid intuitive eye
calculations based on the ball's direction, speed, and trajectory. Since
physics is partly concerned with velocities, trajectories, and colli-
sions of objects, much of the math taught to prepare a student for

physics deals with relationships and formulas that can be used to express motion and acceleration. A young woman who has not closely observed objects travel and collide cannot appreciate the power of mathematics. . . .

Conclusion

42 After surveying the summaries of research in this area and interviewing people who claim to be incompetent at mathematics, I have reached a conclusion. Apart from general intelligence, which is probably equally distributed among males and females, the most important elements in predicting success at learning math are motivation, temperament, attitude, and interest. These are at least as salient as genes and hormones (about which we really know very little in relation to math), "innate reasoning ability" (about which there is much difference of opinion), or number sense. This does not, however, mean that there are no sex differences at all.

43 What is ironic (and unexpected) is that as far as I can judge sex differences seem to be lodged in *acquired skills;* not in computation, visualization, and reasoning *per se,* but in ability to take a math problem apart, in willingness to tolerate certain kinds of ambiguity, and in careful attention to mathematical detail. Such temperamental characteristics as persistence and willingness to take risks may be as important in doing math as pure memory or logic. And attitude and self-image, particularly during adolescence when the pressures to conform are at their greatest, may be even more important than temperament. Negative attitudes, as we all know from personal experience, can powerfully inhibit intellect and curiosity and can keep us from learning what is well within our power to understand. . . .

STUDY QUESTIONS

1. Early in the essay, Tobias cites the results of a number of studies showing that girls do less well at mathematics than

boys. Are the studies diverse and numerous enough to support such a generalization?

2. How many explanations does Tobias consider to account for the studies showing that girls perform less well than boys at math?

3. On what grounds does Tobias reject the view that biological differences account for the differences in performance? Does she consider any evidence in favor of the biological explanation?

4. Toward the end of the essay, how does Tobias link activities such as playing with dolls to poorer performances in mathematics? What skills does playing with dolls develop? Are any of them related to the skills needed for mathematics?

5. Do you suffer from "math anxiety"? If so, has Tobias convinced you that you can, in principle, do well at mathematics? Why or why not?

Reference Chapters: 7, 15, 18

E. D. HIRSCH, JR.

Literacy and Cultural Literacy

E. D. Hirsch, Jr., is the William R. Kenan Professor of English at the University of Virginia, Charlottesville. In the fol-

[E. D. Hirsch, Jr., excerpt from Chapter 1 of *Cultural Literacy*. Boston: Houghton Mifflin, 1987, pp. 3–13, references deleted.]

lowing selection, taken from his 1987 best seller, Cultural Literacy, *Hirsch argues that literacy has declined and that this has grave cultural implications.*

1   Consider the implications of the following experiment described in an article in *Scientific American.* A researcher goes to Harvard Square in Cambridge, Massachusetts, with a tape recorder hidden in his coat pocket. Putting a copy of the *Boston Globe* under his arm, he pretends to be a native. He says to passers-by, "How do you get to Central Square?" The passers-by, thinking they are addressing a fellow Bostonian, don't even break their stride when they give their replies, which consist of a few words like "First stop on the subway."

2   The next day the researcher goes to the same spot, but this time he presents himself as a tourist, obviously unfamiliar with the city. "I'm from out of town," he says. "Can you tell me how to get to Central Square?" This time the tapes show that people's answers are much longer and more rudimentary. A typical one goes, "Yes, well you go down on the subway. You can see the entrance over there, and when you get downstairs you buy a token, put it in the slot, and you go over to the side that says Quincy. You take the train headed for Quincy, but you get off very soon, just the first stop is Central Square, and be sure you get off there. You'll know it because there's a big sign on the wall. It says Central Square." And so on.

3   Passers-by were intuitively aware that communication between strangers requires an estimate of how much relevant information can be taken for granted in the other person. If they can take a lot for granted, their communications can be short and efficient, subtle and complex. But if strangers share very little knowledge, their communications must be long and relatively rudimentary.

4   In order to put in perspective the importance of background knowledge in language, I want to connect the lack of it with our recent lack of success in teaching mature literacy to all students. The most broadly based evidence about our teaching of literacy comes from the National Assessment of Educational Progress (NAEP). This nationwide measurement, mandated by Congress, shows that between 1970 and 1980 seventeen-year-olds declined in their ability to understand written materials, and the decline was especially striking in the top group, those able to read at an "advanced" level.

Although these scores have now begun to rise, they remain alarmingly low. Still more precise quantitative data have come from the scores of the verbal Scholastic Aptitude Test (SAT). According to John B. Carroll, a distinguished psychometrician, the verbal SAT is essentially a test of "advanced vocabulary knowledge," which makes it a fairly sensitive instrument for measuring levels of literacy. It is well known that verbal SAT scores have declined dramatically in the past fifteen years, and though recent reports have shown them rising again, it is from a very low base. Moreover, performance on the verbal SAT has been slipping steadily *at the top*. Ever fewer numbers of our best and brightest students are making high scores on the test.

Before the College Board disclosed the full statistics in 1984, antialarmists could argue that the fall in average verbal scores could be explained by the rise in the number of disadvantaged students taking the SATs. That argument can no longer be made. It's now clear that not only our disadvantaged but also our best educated and most talented young people are showing diminished verbal skills. To be precise, out of a constant pool of about a million test takers each year, 56 percent more students scored above 600 in 1972 than did so in 1984. More startling yet, the percentage drop was even greater for those scoring above 650—73 percent.

In the mid 1980s American business leaders have become alarmed by the lack of communication skills in the young people they employ. Recently, top executives of some large U.S. companies, including CBS and Exxon, met to discuss the fact that their younger middle-level executives could no longer communicate their ideas effectively in speech or writing. This group of companies has made a grant to the American Academy of Arts and Sciences to analyze the causes of this growing problem. They want to know why, despite breathtaking advances in the technology of communication, the effectiveness of business communication has been slipping, to the detriment of our competitiveness in the world. The figures from NAEP surveys and the scores on the verbal SAT are solid evidence that literacy has been declining in this country just when our need for effective literacy has been sharply rising.

I now want to juxtapose some evidence for another kind of educational decline, one that is related to the drop in literacy.

During the period 1970–1985, the amount of shared knowledge that we have been able to take for granted in communicating with our fellow citizens has also been declining. More and more of our young people don't know things we used to assume they knew.

8   A side effect of the diminution in shared information has been a noticeable increase in the number of articles in such publications as *Newsweek* and the *Wall Street Journal* about the surprising ignorance of the young. My son John, who recently taught Latin in high school and eighth grade, often told me of experiences which indicate that these articles are not exaggerated. In one of his classes he mentioned to his students that Latin, the language they were studying, is a dead language that is no longer spoken. After his pupils had struggled for several weeks with Latin grammar and vocabulary, this news was hard for some of them to accept. One girl raised her hand to challenge my son's claim. "What do they speak in Latin America?" she demanded.

9   At least she had heard of Latin America. Another day my son asked his Latin class if they knew the name of an epic poem by Homer. One pupil shot up his hand and eagerly said, "The Alamo!" Was it just a slip for *The Iliad?* No, he didn't know what the Alamo was, either. To judge from other stories about information gaps in the young, many American schoolchildren are less well informed than this pupil. The following, by Benjamin J. Stein, is an excerpt from one of the most evocative recent accounts of youthful ignorance.

> I spend a lot of time with teen agers. Besides employing three of them part-time, I frequently conduct focus groups at Los Angeles area high schools to learn about teen agers' attitudes towards movies or television shows or nuclear arms or politicians. . . .
>
> I have not yet found one single student in Los Angeles, in either college or high school, who could tell me the years when World War II was fought. Nor have I found one who could tell me the years when World War I was fought. Nor have I found one who knew when the American Civil War was fought. . . .
>
> A few have known how many U.S. senators California has, but none has known how many Nevada or Oregon has. ("Really? Even though they're so small?") . . . Only two could tell me where Chicago is, even in the vaguest terms. (My particular favorite geography

lesson was the junior at the University of California at Los Angeles who thought that Toronto must be in Italy. My second-favorite geography lesson is the junior at USC, a pre-law student, who thought that Washington, D.C., was in Washington State.) . . .

Only two could even approximately identify Thomas Jefferson. Only one could place the date of the Declaration of Independence. None could name even one of the first ten amendments to the Constitution or connect them with the Bill of Rights. . . .

On and on it went. On and on it goes. I have mixed up episodes of ignorance of facts with ignorance of concepts because it seems to me that there is a connection. . . . The kids I saw (and there may be lots of others who are different) are not mentally prepared to continue the society because they basically do not understand the society well enough to value it.

My son assures me that his pupils are not ignorant. They know   10
a great deal. Like every other human group they share a tremendous amount of knowledge among themselves, much of it learned in school. The trouble is that, from the standpoint of their literacy and their ability to communicate with others in our culture, what they know is ephemeral and narrowly confined to their own generation. Many young people strikingly lack the information that writers of American books and newspapers have traditionally taken for granted among their readers from all generations. For reasons explained in this book, our children's lack of intergenerational information is a serious problem for the nation. The decline of literacy and the decline of shared knowledge are closely related, interdependent facts.

The evidence for the decline of shared knowledge is not just   11
anecdotal. In 1978 NAEP issued a report which analyzed a large quantity of data showing that our children's knowledge of American civics had dropped significantly between 1969 and 1976. The performance of thirteen-year-olds had dropped an alarming 11 percentage points. That the drop has continued since 1976 was confirmed by preliminary results from a NAEP study conducted in late 1985. It was undertaken both because of concern about declining knowledge and because of the growing evidence of a causal connection between the drop in shared information and in literacy. The Foundations of Literacy project is measuring some of the specific

information about history and literature that American seventeen-year-olds possess.

12   Although the full report will not be published until 1987, the preliminary field tests are disturbing. If these samplings hold up, and there is no reason to think they will not, then the results we will be reading in 1987 will show that two thirds of our seventeen-year-olds do not know that the Civil War occurred between 1850 and 1900. Three quarters do not know what *reconstruction* means. Half do not know the meaning of *Brown decision* and cannot identify either Stalin or Churchill. Three quarters are unfamiliar with the names of standard American and British authors. Moreover, our seventeen-year-olds have little sense of geography or the relative chronology of major events. Reports of youthful ignorance can no longer be considered merely impressionistic.

13   My encounter in the seventies with this widening knowledge gap first caused me to recognize the connection between specific background knowledge and mature literacy. The research I was doing on the reading and writing abilities of college students made me realize two things. First, we cannot assume that young people today know things that were known in the past by almost every literate person in the culture. For instance, in one experiment conducted in Richmond, Virginia, our seventeen- and eighteen-year-old subjects did not know who Grant and Lee were. Second, our results caused me to realize that we cannot treat reading and writing as empty skills, independent of specific knowledge. The reading skill of a person may vary greatly from task to task. The level of literacy exhibited in each task depends on the relevant background information that the person possesses.

14   The lack of wide-ranging background information among young men and women now in their twenties and thirties is an important cause of the illiteracy that large corporations are finding in their middle-level executives. In former days, when business people wrote and spoke to one another, they could be confident that they and their colleagues had studied many similar things in school. They could talk to one another with an efficiency similar to that of native Bostonians who speak to each other in the streets of Cambridge. But today's high school graduates do not reliably share much common information, even when they graduate from the same

school. If young people meet as strangers, their communications resemble the uncertain, rudimentary explanations recorded in the second part of the Cambridge experiment.

My father used to write business letters that alluded to Shake-   15
speare. These allusions were effective for conveying complex messages to his associates, because, in his day, business people could make such allusions with every expectation of being understood. For instance, in my father's commodity business, the timing of sales and purchases was all-important, and he would sometimes write or say to his colleagues, "There is a tide," without further elaboration. Those four words carried not only a lot of complex information, but also the persuasive force of a proverb. In addition to the basic practical meaning, "Act now!" what came across was a lot of implicit reasons why immediate action was important.

For some of my younger readers who may not recognize the   16
allusion, the passage from *Julius Caesar* is:

> There is a tide in the affairs of men
> Which taken at the flood leads on to fortune;
> Omitted, all the voyage of their life
> Is bound in shallows and in miseries.
> On such a full sea are we now afloat,
> And we must take the current when it serves,
> Or lose our ventures.

To say "There is a tide" is better than saying "Buy (or sell) now and you'll cover expenses for the whole year, but if you fail to act right away, you may regret it the rest of your life." That would be twenty-seven words instead of four, and while the bare message of the longer statement would be conveyed, the persuasive force wouldn't. Think of the demands of such a business communication. To persuade somebody that your recommendation is wise and well-founded, you have to give lots of reasons and cite known examples and authorities. My father accomplished that and more in four words, which made quoting Shakespeare as effective as any efficiency consultant could wish. The moral of this tale is not that reading Shakespeare will help one rise in the business world. My point is a broader one. The fact that middle-level executives no longer share literate background knowledge is a chief cause of their inability to communicate effectively.

The Nature and Use of Cultural Literacy

17   The documented decline in shared knowledge carries implications that go far beyond the shortcomings of executives and extend to larger questions of educational policy and social justice in our country. Mina Shaughnessy was a great English teacher who devoted her professional life to helping disadvantaged students become literate. At the 1980 conference dedicated to her memory, one of the speakers who followed me to the podium was the Harvard historian and sociologist Orlando Patterson. To my delight he departed from his prepared talk to mention mine. He seconded my argument that shared information is a necessary background to true literacy. Then he extended and deepened the ideas I had presented. Here is what Professor Patterson said, as recorded in the *Proceedings* of the conference.

> Industrialized civilization [imposes] a growing cultural and structural complexity which requires persons to have a broad grasp of what Professor Hirsch has called cultural literacy: a deep understanding of mainstream culture, which no longer has much to do with white Anglo-Saxon Protestants, but with the imperatives of industrial civilization. It is the need for cultural literacy, a profound conception of the whole civilization, which is often neglected in talk about literacy.

Patterson continued by drawing a connection between background information and the ability to hold positions of responsibility and power. He was particularly concerned with the importance for blacks and other minorities of possessing this information, which is essential for improving their social and economic status.

> The people who run society at the macro-level must be literate in this culture. For this reason, it is dangerous to overemphasize the problems of basic literacy or the relevancy of literacy to specific tasks, and more constructive to emphasize that blacks will be condemned in perpetuity to oversimplified, low-level tasks and will never gain their rightful place in controlling the levers of power unless they also acquire literacy in this wider cultural sense.

Although Patterson focused his remarks on the importance of cultural literacy for minorities, his observations hold for every culturally illiterate person in our nation. Indeed, as he observed, cultural literacy is not the property of any group or class.

> To assume that this wider culture is static is an error; in fact it is not. It's not a WASP culture; it doesn't belong to any group. It is essentially and constantly changing, and it is open. What is needed is recognition that the accurate metaphor or model for this wider literacy is not domination, but dialectic; each group participates and contributes, transforms and is transformed, as much as any other group. . . . The English language no longer belongs to any single group or nation. The same goes for any other area of the wider culture.

As Professor Patterson suggested, being taught to decode elementary reading materials and specific, job-related texts cannot constitute true literacy. Such basic training does not make a person literate with respect to newspapers or other writings addressed to a general public. Moreover, a directly practical drawback of such narrow training is that it does not prepare anyone for technological change. Narrow vocational training in one state of a technology will not enable a person to read manuals that explain new developments in the same technology. In modern life we need general knowledge that enables us to deal with new ideas, events, and challenges. In today's world, general cultural literacy is more useful than what Professor Patterson terms "literacy to a specific task," because general literate information is the basis for many changing tasks. 18

Cultural literacy is even more important in the social sphere. The aim of universal literacy has never been a socially neutral mission in our country. Our traditional social goals were unforgettably renewed for us by Martin Luther King, Jr., in his "I Have a Dream" speech. King envisioned a country where the children of former slaves sit down at the table of equality with the children of former slave owners, where men and women deal with each other as equals and judge each other on their characters and achievements rather than their origins. Like Thomas Jefferson, he had a dream of a society founded not on race or class but on personal merit. 19

In the present day, that dream depends on mature literacy. No modern society can hope to become a just society without a high 20 level of universal literacy. Putting aside for the moment the practical arguments about the economic uses of literacy, we can contemplate the even more basic principle that underlies our national system of education in the first place—that people in a democracy can be entrusted to decide all important matters for themselves because they can deliberate and communicate with one another. Universal literacy is inseparable from democracy and is the canvas for Martin Luther King's picture as well as for Thomas Jefferson's.

21 Both of these leaders understood that just having the right to vote is meaningless if a citizen is disenfranchised by illiteracy or semiliteracy. Illiterate and semiliterate Americans are condemned not only to poverty, but also to the powerlessness of incomprehension. Knowing that they do not understand the issues, and feeling prey to manipulative oversimplifications, they do not trust the system of which they are supposed to be the masters. They do not feel themselves to be active participants in our republic, and they often do not turn out to vote. The civic importance of cultural literacy lies in the fact that true enfranchisement depends upon knowledge, knowledge upon literacy, and literacy upon cultural literacy.

22 To be truly literate, citizens must be able to grasp the meaning of any piece of writing addressed to the general reader. All citizens should be able, for instance, to read newspapers of substance, about which Jefferson made the following famous remark:

> Were it left to me to decide whether we should have a government without newspapers, or newspapers without a government, I should not hesitate a moment to prefer the latter.
>
> But I should mean that every man should receive those papers and be capable of reading them.

Jefferson's last comment is often omitted when the passage is quoted, but it's the crucial one.

STUDY QUESTIONS

1. What is the point of the experiment from *Scientific American* that Hirsch reports in paragraphs 1 and 2? Does Hirsch use it as part of an argument?

2. Does Hirsch define "cultural literacy"?
3. How many pieces of evidence does Hirsch present to justify his claim that cultural literacy is declining? Are his sources of evidence comprehensive enough to justify his claim?
4. Does Hirsch discuss the causes of the decline of literacy?
5. How many important values does Hirsch say depend on cultural literacy? For each value, how does Hirsch argue that it depends on cultural literacy?

Reference Chapters: 3, 15

CAROLINE BIRD

College Is a Waste of Time and Money

*Caroline Bird was born in New York in 1915. She was edu-
cated at Vassar College and the University of Toledo, and is
the author of* The Crowding Syndrome *and* The Case
Against College, *from which the following selection is excerpted. As the title of the essay indicates, Bird argues that college is not the good investment it is usually thought to be.*

A great majority of our nine million college students are not in   1
school because they want to be or because they want to learn. They
are there because it has become the thing to do or because college
is a pleasant place to be; because it's the only way they can get

[Caroline Bird, excerpt from *The Case Against College*. New York: David
McKay Co., 1975, pp. 281–289.]

parents or taxpayers to support them without working at a job they don't like; because Mother wanted them to go, or some other reason entirely irrelevant to the course of studies for which college is supposedly organized.

2   As I crisscross the United States lecturing on college campuses, I am dismayed to find that professors and administrators, when pressed for a candid opinion, estimate that no more than 25 percent of their students are turned on by classwork. For the rest, college is at best a social center or aging vat, and at worst a young folks' home or even a prison that keeps them out of the mainstream of economic life for a few more years.

3   The premise—which I no longer accept—that college is the best place for all high-school graduates grew out of a noble American ideal. Just as the United States was the first nation to aspire to teach every small child to read and write, so, during the 1950s, we became the first and only great nation to aspire to higher education for all. During the '60s we damned the expense and built great state university systems as fast as we could. And adults—parents, employers, high-school counselors—began to push, shove and cajole youngsters to "get an education."

4   It became a mammoth industry, with taxpayers footing more than half the bill. By 1970, colleges and universities were spending more than 30 billion dollars annually. But still only half our high-school graduates were going on. According to estimates made by the economist Fritz Machlup, if we had been educating every young person until age 22 in that year of 1970, the bill for higher education would have reached 47.5 billion dollars, 12.5 billion more than the total corporate profits for the year.

5   Figures such as these have begun to make higher education for all look financially prohibitive, particularly now when colleges are squeezed by the pressures of inflation and a drop-off in the growth of their traditional market.

6   Predictable demography has caught up with the university empire builders. Now that the record crop of postwar babies has graduated from college, the rate of growth of the student population has begun to decline. To keep their mammoth plants financially solvent, many institutions have begun to use hard-sell, Madison-Avenue techniques to attract students. They sell college like soap, promoting features they think students want: innovative programs,

an environment conducive to meaningful personal relationships, and a curriculum so free that it doesn't sound like college at all.

Pleasing the customers is something new for college administrators. Colleges have always known that most students don't like to study, and that at least part of the time they are ambivalent about college, but before the student riots of the 1960s educators never thought it either right or necessary to pay any attention to student feelings. But when students rebelling against the Vietnam war and the draft discovered they could disrupt a campus completely, administrators had to act on some student complaints. Few understood that the protests had tapped the basic discontent with college itself, a discontent that did not go away when the riots subsided. 7

Today students protest individually rather than in concert. They turn inward and withdraw from active participation. They drop out to travel to India or to feed themselves on subsistence farms. Some refuse to go to college at all. Most, of course, have neither the funds nor the self-confidence for constructive articulation of their discontent. They simply hang around college unhappily and reluctantly. 8

All across the country, I have been overwhelmed by the prevailing sadness on American campuses. Too many young people speak little, and then only in drowned voices. Sometimes the mood surfaces as diffidence, wariness, or coolness, but whatever its form, it looks like a defense mechanism, and that rings a bell. This is the way it used to be with women, and just as society had systematically damaged women by insisting that their proper place was in the home, so we may be systematically damaging 18-year-olds by insisting that their proper place is in college. 9

Campus watchers everywhere know what I mean when I say students are sad, but they don't agree on the reason for it. During the Vietnam war some ascribed the sadness to the draft; now others blame affluence, or say it has something to do with permissive upbringing. 10

Not satisfied with any of these explanations, I looked for some answers with the journalistic tools of my trade—scholarly studies, economic analyses, the historical record, the opinions of the especially knowledgeable, conversations with parents, professors, college administrators, and employers, all of whom spoke as alumni too. Mostly I learned from my interviews with hundreds of young people on and off campuses all over the country. 11

12   My unnerving conclusion is that students are sad because they are not needed. Somewhere between the nursery and the employment office, they become unwanted adults. No one has anything in particular against them. But no one knows what to do with them either. We already have too many people in the world of the 1970s, and there is no room for so many newly minted 18-year-olds. So we temporarily get them out of the way by sending them to college where in fact only a few belong.

13   To make it more palatable, we fool ourselves into believing that we are sending them there for their own best interests, and that it's good for them, like spinach. Some, of course, learn to like it, but most wind up preferring green peas.

14   Educators admit as much. Nevitt Sanford, distinguished student of higher education, says students feel they are "capitulating to a kind of voluntary servitude." Some of them talk about their time in college as if it were a sentence to be served. I listened to a 1970 Mount Holyoke graduate: "For two years I was really interested in science, but in my junior and senior years I just kept saying, 'I've done two years; I'm going to finish.' When I got out I made up my mind that I wasn't going to school anymore because so many of my courses had been bullshit."

15   But bad as it is, college is often preferable to a far worse fate. It is better than the drudgery of an uninspiring nine-to-five job, and better than doing nothing when no jobs are available. For some young people, it is a graceful way to get away from home and become independent without losing the financial support of their parents. And sometimes it is the only alternative to an intolerable home situation.

16   It is difficult to assess how many students are in college reluctantly. The conservative Carnegie Commission estimates from 5 to 30 percent. Sol Linowitz, who was once chairman of a special committee on campus tension of the American Council on Education, found that "a significant number were not happy with their college experience because they felt they were there only in order to get the 'ticket to the big show' rather than to spend the years as productively as they otherwise could."

17   Older alumni will identify with Richard Baloga, a policeman's son, who stayed in school even though he "hated it" because he

thought it would do him some good. But fewer students each year feel this way. Daniel Yankelovich has surveyed undergraduate attitudes for a number of years, and reported in 1971 that 74 percent thought education was "very important." But just two years earlier, 80 percent thought so.

The doubters don't mind speaking up. Leon Lefkowitz, chairman   18
of the department of social studies at Central High School in Valley Stream, New York, interviewed 300 college students at random, and reports that 200 of them didn't think that the education they were getting was worth the effort. "In two years I'll pick up a diploma," said one student, "and I can honestly say it was a waste of my father's bread."

Nowadays, says one sociologist, you don't have to have a reason   19
for going to college; it's an institution. His definition of an institution is an arrangement everyone accepts without question; the burden of proof is not on why you go, but why anyone thinks there might be a reason for not going. The implication is that an 18-year-old is too young and confused to know what he wants to do, and that he should listen to those who know best and go to college.

I don't agree. I believe that college has to be judged not on what   20
other people think is good for students, but on how good it feels to the students themselves.

I believe that people have an inside view of what's good for them.   21
If a child doesn't want to go to school some morning, better let him stay at home, at least until you find out why. Maybe he knows something you don't. It's the same with college. If high-school graduates don't want to go, or if they don't want to go right away, they may perceive more clearly than their elders that college is not for them. It is no longer obvious that adolescents are best off studying a core curriculum that was constructed when all educated men could agree on what made them educated, or that professors, advisors, or parents can be of any particular help to young people in choosing a major or a career. High-school graduates see college graduates driving cabs, and decide it's not worth going. College students find no intellectual stimulation in their studies and drop out.

If students believe that college isn't necessarily good for them,   22
you can't expect them to stay on for the general good of mankind.

They don't go to school to beat the Russians to Jupiter, improve the national defense, increase the GNP, or create a new market for the arts—to mention some of the benefits taxpayers are supposed to get for supporting higher education.

23   Nor should we expect to bring about social equality by putting all young people through four years of academic rigor. At best, it's a roundabout and expensive way to narrow the gap between the highest and lowest in our society anyway. At worst, it is unconsciously elitist. Equalizing opportunity through universal higher education subjects the whole population to the intellectual mode natural only to a few. It violates the fundamental egalitarian principle of respect for the differences between people.

24   Of course, most parents aren't thinking of the "higher" good at all. They send their children to college because they are convinced young people benefit financially from those four years of higher education. But if money is the only goal, college is the dumbest investment you can make. I say this because a young banker in Poughkeepsie, New York, Stephen G. Necel, used a computer to compare college as an investment with other investments available in 1974 and college did not come out on top.

25   For the sake of argument, the two of us invented a young man whose rich uncle gave him, in cold cash, the cost of a four-year education at any college he chose, but the young man didn't have to spend the money on college. After bales of computer paper, we had our mythical student write to his uncle: "Since you said I could spend the money foolishly if I wished, I am going to blow it all on Princeton."

26   The much respected financial columnist Sylvia Porter echoed the common assumption when she said last year, "A college education is among the very best investments you can make in your entire life." But the truth is not quite so rosy, even if we assume that the Census Bureau is correct when it says that as of 1972, a man who completed four years of college would expect to earn $199,000 more between the ages of 22 and 64 than a man who had only a high-school diploma.

27   If a 1972 Princeton-bound high-school graduate had put the $34,181 that his four years of college would have cost him into a savings bank at 7.5 percent interest compounded daily, he would

have had at age 64 a total of $1,129,200, or $528,200 more than the earnings of a male college graduate, and more than five times as much as the $199,000 extra the more educated man could expect to earn between 22 and 64.

The big advantage of getting your college money in cash now is that you can invest it in something that has a higher return than a diploma. For instance, a Princeton-bound high-school graduate of 1972 who liked fooling around with cars could have banked his $34,181, and gone to work at the local garage at close to $1,000 more per year than the average high-school graduate. Meanwhile, as he was learning to be an expert auto mechanic, his money would be ticking away in the bank. When he became 28, he would have earned $7,199 less on his job from age 22 to 28 than his college-educated friend, but he would have had $73,113 in his passbook—enough to buy out his boss, go into the used-car business, or acquire his own new-car dealership. If successful in business, he could expect to make more than the average college graduate. And if he had the brains to get into Princeton, he would be just as likely to make money without the four years spent on campus. Unfortunately, few college-bound high-school graduates get the opportunity to bank such a large sum of money, and then wait for it to make them rich. And few parents are sophisticated enough to understand that in financial returns alone, their children would be better off with the money than with the education.

28

Rates of return and dollar signs on education are fascinating brain teasers, but obviously there is a certain unreality to the game. Quite aside from the noneconomic benefits of college, and these should loom larger once the dollars are cleared away, there are grave difficulties in assigning a dollar value to college at all.

29

In fact there is no real evidence that the higher income of college graduates is due to college. College may simply attract people who are slated to earn more money anyway; those with higher IQs, better family backgrounds, a more enterprising temperament. No one who has wrestled with the problem is prepared to attribute all of the higher income to the impact of college itself.

30

Christopher Jencks, author of *Inequality,* a book that assesses the effect of family and schooling in America, believes that education in general accounts for less than half of the difference in income in

31

the American population. "The biggest single source of income differences," writes Jencks, "seems to be the fact that men from high-status families have higher incomes than men from low-status families even when they enter the same occupations, have the same amount of education, and have the same test scores."

32   Jacob Mincer of the National Bureau of Economic Research and Columbia University states flatly that of "20 to 30 percent of students at any level, the additional schooling has been a waste, at least in terms of earnings." College fails to work its income-raising magic for almost a third of those who go. More than half of those people in 1972 who earned $15,000 or more reached that comfortable bracket without the benefit of a college diploma. Jencks says that financial success in the U.S. depends a good deal on luck, and the most sophisticated regression analyses have yet to demonstrate otherwise.

33   But most of today's students don't go to college to earn more money anyway. In 1968, when jobs were easy to get, Daniel Yankelovich made his first nationwide survey of students. Sixty-five percent of them said they "would welcome less emphasis on money." By 1973, when jobs were scarce, that figure jumped to 80 percent.

34   The young are not alone. Americans today are all looking less to the pay of a job than to the work itself. They want "interesting" work that permits them "to make a contribution," express themselves" and "use their special abilities," and they think college will help them find it.

35   Jerry Darring of Indianapolis knows what it is to make a dollar. He worked with his father in the family plumbing business, on the line at Chevrolet, and in the Chrysler foundry. He quit these jobs to enter Wright State University in Dayton, Ohio, because "in a job like that a person only has time to work, and after that he's so tired that he can't do anything else but come home and go to sleep."

36   Jerry came to college to find work "helping people." And he is perfectly willing to spend the dollars he earns at dull, well-paid work to prepare for lower-paid work that offers the reward of service to others.

37   Jerry's case is not unusual. No one works for money alone. In order to deal with the nonmonetary rewards of work, economists

have coined the concept of "psychic income" which according to one economic dictionary means "income that is reckoned in terms of pleasure, satisfaction, or general feelings of euphoria."

Psychic income is primarily what college students mean when they talk about getting a good job. During the most affluent years of the late 1960s and early 1970s college students told their placement officers that they wanted to be researchers, college professors, artists, city planners, social workers, poets, book publishers, archeologists, ballet dancers, or authors.

The psychic income of these and other occupations popular with students is so high that these jobs can be filled without offering high salaries. According to one study, 93 percent of urban university professors would choose the same vocation again if they had the chance, compared with only 16 percent of unskilled auto workers. Even though the monetary gap between college professor and auto worker is now surprisingly small, the difference in psychic income is enormous.

But colleges fail to warn students that jobs of these kinds are hard to come by, even for qualified applicants, and they rarely accept the responsibility of helping students choose a career that will lead to a job. When a young person says he is interested in helping people, his counselor tells him to become a psychologist. But jobs in psychology are scarce. The Department of Labor, for instance, estimates there will be 4,300 new jobs for psychologists in 1975 while colleges are expected to turn out 58,430 B.A.s in psychology that year.

Of 30 psych majors who reported back to Vassar what they were doing a year after graduation in 1972, only five had jobs in which they could possibly use their courses in psychology, and two of these were working for Vassar.

The outlook isn't much better for students majoring in other psychic-pay disciplines: sociology, English, journalism, anthropology, forestry, education. Whatever college graduates want to do, most of them are going to wind up doing what there is to do.

John Shingleton, director of placement at Michigan State University, accuses the academic community of outright hypocrisy. "Educators have never said, 'Go to college and get a good job,' but this has been implied, and now students expect it. . . . If we care what happens to students after college, then let's get involved with what should be one of the basic purposes of education: career preparation."

44   In the 1970s, some of the more practical professors began to see that jobs for graduates meant jobs for professors too. Meanwhile, students themselves reacted to the shrinking job market, and a "new vocationalism" exploded on campus. The press welcomed the change as a return to the ethic of achievement and service. Students were still idealistic, the reporters wrote, but they now saw that they could best make the world better by healing the sick as physicians or righting individual wrongs as lawyers.

45   But there are no guarantees in these professions either. The American Enterprise Institute estimated in 1971 that there would be more than the target ratio of 100 doctors for every 100,000 people in the population by 1980. And the odds are little better for would-be lawyers. Law schools are already graduating twice as many new lawyers every year as the Department of Labor thinks will be needed, and the oversupply is growing every year.

46   And it's not at all apparent that what is actually learned in a "professional" education is necessary for success. Teachers, engineers and others I talked to said they find that on the job they rarely use what they learned in school. In order to see how well college prepared engineers and scientists for actual paid work in their fields, The Carnegie Commission queried all the employees with degrees in these fields in two large firms. Only one in five said the work they were doing bore a "very close relationship" to their college studies, while almost a third saw "very little relationship at all." An overwhelming majority could think of many people who were doing their same work, but had majored in different fields.

47   Majors in nontechnical fields report even less relationship between their studies and their jobs. Charles Lawrence, a communications major in college and now the producer of "Kennedy & Co.," the Chicago morning television show, says, "You have to learn all that stuff and you never use it again. I learned my job doing it." Others employed as architects, nurses, teachers and other members of the so-called learned professions report the same thing.

48   Most college administrators admit that they don't prepare their graduates for the job market. "I just wish I had the guts to tell parents that when you get out of this place you aren't prepared to

do anything," the academic head of a famous liberal-arts college told us. Fortunately, for him, most people believe that you don't have to defend a liberal-arts education on those grounds. A liberal-arts education is supposed to provide you with a value system, a standard, a set of ideas, not a job. "Like Christianity, the liberal arts are seldom practiced and would probably be hated by the majority of the populace if they were," said one defender.

The analogy is apt. The fact is, of course, that the liberal arts are   49
a religion in every sense of that term. When people talk about them, their language becomes elevated, metaphorical, extravagant, theoretical and reverent. And faith in personal salvation by the liberal arts is professed in a creed intoned on ceremonial occasions such as commencements.

If the liberal arts are a religious faith, the professors are its priests.   50
But disseminating ideas in a four-year college curriculum is slow and most expensive. If you want to learn about Milton, Camus, or even Margaret Mead, you can find them in paperback books, the public library, and even on television.

And when most people talk about the value of a college educa-   51
tion, they are not talking about great books. When at Harvard commencement, the president welcomes the new graduates into "the fellowship of educated men and women," what he could be saying is, "Here is a piece of paper that is a passport to jobs, power and instant prestige." As Glenn Bassett, a personnel specialist at G.E., says, "In some parts of G.E., a college degree appears completely irrelevant to selection to, say, a manager's job. In most, however, it is a ticket of admission."

But now that we have doubled the number of young people   52
attending college, a diploma cannot guarantee even that. The most charitable conclusion we can reach is that college probably has very little, if any, effect on people and things at all. Today, the false premises are easy to see:

First, college doesn't make people intelligent, ambitious, happy,   53
or liberal. It's the other way around. Intelligent, ambitious, happy, liberal people are attracted to higher education in the first place.

Second, college can't claim much credit for the learning experi-   54
ences that really change students while they are there. Jobs, friends, history, and most of all the sheer passage of time, have as big an impact as anything even indirectly related to the campus.

55   Third, colleges have changed so radically that a freshman entering in the fall of 1974 can't be sure to gain even the limited value
research studies assigned to colleges in the '60s. The sheer size of
undergraduate campuses of the 1970s makes college even less
stimulating now than it was 10 years ago. Today even motivated
students are disappointed with their college courses and professors.

56   Finally, a college diploma no longer opens as many vocational
doors. Employers are beginning to realize that when they pay extra
for someone with a diploma, they are paying only for an empty
credential. The fact is that most of the work for which employers
now expect college training is now or has been capably done in the
past by people without higher educations.

57   College, then, may be a good place for those few young people
who are really drawn to academic work, who would rather read than
eat, but it has become too expensive, in money, time, and intellectual effort to serve as a holding pen for large numbers of our young.
We ought to make it possible for those reluctant, unhappy students
to find alternative ways of growing up, and more realistic preparation for the years ahead.

Study Questions

1. Judging from your own experience, do you agree with Bird's
   claim that there is a "prevailing sadness" among college
   students? If you agree, do you think the sadness is for the
   reasons Bird suggests?
2. How many good reasons can you think of for going to college?
   Of these reasons, how many does Bird consider and reject as
   making college worthwhile? Does it make any difference to
   her argument if one has a specific career goal in mind?
3. How should one go about deciding whether a college degree
   is a good financial investment? What are all of the relevant
   factors? How does one factor in "psychic income"? (See paragraph 37 for a definition of "psychic income.")
4. In paragraphs 25–33, Bird presents her argument that a

college degree is not a good financial investment. Has Bird taken into account all of the relevant factors?

5. What skills are necessary to succeed at most jobs? Why, according to paragraphs 46–56, does Bird think that students do not acquire the necessary skills at college?

Reference Chapters: 5, 7

PART SEVEN

On Justice and Rights

NICCOLÒ MACHIAVELLI

On Cruelty and Clemency: Whether It Is Better to Be Loved or Feared

Niccolò Machiavelli (1469–1527) was a Florentine statesman and author of Il Principe *(The Prince, 1513), from which the following selection is excerpted.* Il Principe *has become a famous and influential work in political theory for its argument that politicians may use dishonesty and other treacherous methods to maintain and increase their power.*

Continuing now with our list of qualities, let me say that every prince should prefer to be considered merciful rather than cruel, yet

1

[Niccolò Machiavelli, "On Cruelty and Clemency: Whether It Is Better to be Loved or Feared," Section XVII of *The Prince*, transl. Robert M. Adams. New York: W. W. Norton, 1977, pp. 47–49, some footnotes deleted.]

he should be careful not to mismanage this clemency of his. People thought Cesare Borgia was cruel, but that cruelty of his reorganized the Romagna, united it, and established it in peace and loyalty. Anyone who views the matter realistically will see that this prince was much more merciful than the people of Florence, who, to avoid the reputation of cruelty, allowed Pistoia to be destroyed. Thus, no prince should mind being called cruel for what he does to keep his subjects united and loyal; he may make examples of a very few, but he will be more merciful in reality than those who, in their tender-heartedness, allow disorders to occur, with their attendant murders and lootings. Such turbulence brings harm to an entire community, while the executions ordered by a prince affect only one individual at a time. A new prince, above all others, cannot possibly avoid a name for cruelty, since new states are always in danger. And Virgil, speaking through the mouth of Dido, says:

> Res dura et regni novitas me talia cogunt
> Moliri, et late fines custode tueri.[1]

Yet a prince should be slow to believe rumors and to commit himself to action on the basis of them. He should not be afraid of his own thoughts; he ought to proceed cautiously, moderating his conduct with prudence and humanity, allowing neither overconfidence to make him careless, nor overtimidity to make him intolerable.

2   Here the question arises: is it better to be loved than feared, or vice versa? I don't doubt that every prince would like to be both; but since it is hard to accommodate these qualities, if you have to make a choice, to be feared is much safer than to be loved. For it is a good general rule about men, that they are ungrateful, fickle, liars and deceivers, fearful of danger and greedy for gain. While you serve their welfare, they are all yours, offering their blood, their belong-ings, their lives, and their children's lives, as we noted above—so long as the danger is remote. But when the danger is close at hand, they turn against you. Then, any prince who has relied on their words and has made no other preparations will come to grief; because friendships that are bought at a price, and not with great-ness and nobility of soul, may be paid for but they are not acquired,

---

1. "Harsh pressures and the newness of my reign / Compel me to these steps; I must maintain / My borders against foreign foes. . . ." (*Aeneid*, II, 563–4).

and they cannot be used in time of need. People are less concerned with offending a man who makes himself loved than one who makes himself feared: the reason is that love is a link of obligation which men, because they are rotten, will break any time they think doing so serves their advantage; but fear involves dread of punishment, from which they can never escape.

Still, a prince should make himself feared in such a way that, even if he gets no love, he gets no hate either; because it is perfectly possible to be feared and not hated, and this will be the result if only the prince will keep his hands off the property of his subjects or citizens, and off their women. When he does have to shed blood, he should be sure to have a strong justification and manifest cause; but above all, he should not confiscate people's property, because men are quicker to forget the death of a father than the loss of a patrimony. Besides, pretexts for confiscation are always plentiful; it never fails that a prince who starts living by plunder can find reasons to rob someone else. Excuses for proceeding against someone's life are much rarer and more quickly exhausted.

But a prince at the head of his armies and commanding a multitude of soldiers should not care a bit if he is considered cruel; without such a reputation, he could never hold his army together and ready for action. Among the marvelous deeds of Hannibal, this was prime: that, having an immense army, which included men of many different races and nations, and which he led to battle in distant countries, he never allowed them to fight among themselves or to rise against him, whether his fortune was good or bad. The reason for this could only be his inhuman cruelty, which, along with his countless other talents [*virtù*], made him an object of awe and terror to his soldiers; and without the cruelty, his other qualities [*le altre sua virtù*] would never have sufficed. The historians who pass snap judgments on these matters admire his accomplishments and at the same time condemn the cruelty which was their main cause.

When I say, "His other qualities would never have sufficed," we can see that this is true from the example of Scipio, an outstanding man not only among those of his own time, but in all recorded history; yet his armies revolted in Spain, for no other reason than his excessive leniency in allowing his soldiers more freedom than military discipline permits. Fabius Maximus rebuked him in the senate for this failing, calling him the corrupter of the Roman armies. When a lieutenant of Scipio's plundered the Locrians, he took no action in behalf of the people, and did nothing to discipline that insolent lieutenant; again, this was the result of his easygoing nature. Indeed, when someone in the senate wanted to excuse him on this occasion, he said there are many men who knew better how to avoid error themselves than how to correct error in others. Such a soft temper would in time have tarnished the fame and glory of Scipio, had he brought it to the office of emperor; but as he lived under the control of the senate, this harmful quality of his not only remained hidden but was considered creditable.

6   Returning to the question of being feared or loved, I conclude that since men love at their own inclination but can be made to fear at the inclination of the prince, a shrewd prince will lay his foundations on what is under his own control, not on what is controlled by others. He should simply take pains not to be hated, as I said.

STUDY QUESTIONS

1. In paragraph 1, Machiavelli argues that since new states are always in danger, a new prince "cannot possibly avoid a name for cruelty." What premises are assumed in this argument?
2. What assumptions about human nature does Machiavelli make in advising the prince?
3. Why does Machiavelli think it is dangerous for a prince to rely on the love of the citizens? What account of love does his argument depend upon?
4. How many historical examples does Machiavelli use to justify the use of cruelty over love? Can you think of any counterexamples from history? For example, did George Washington succeed by following Machiavelli's advice?
5. Is the argument in paragraph 6 the same as the argument in paragraph 2?
6. Machiavelli argues that a prince can strive to be obeyed out of love or out of fear. Are these the only options? If not, has Machiavelli presented us with a false alternative?

Reference Chapters: 7, 15

THOMAS JEFFERSON

The Declaration of Independence

Thomas Jefferson (1743–1826) was governor of Virginia, a lawyer, writer, architect, founder of the University of Virginia, and the third President of the United States.

When in the course of human events, it becomes necessary for one people to dissolve the political bands which have connected them with another, and to assume among the Powers of the earth, the separate and equal station to which the Laws of Nature and of Nature's God entitle them, a decent respect to the opinions of mankind requires that they should declare the causes which impel them to the separation.

We hold these truths to be self-evident, that all men are created equal, that they are endowed by their Creator with certain unalienable Rights, that among these are Life, Liberty and the pursuit of Happiness. That to secure these rights, Governments are instituted among Men deriving their just powers from the consent of the governed. That whenever any Form of Government becomes destructive of these ends, it is the Right of the People to alter or to abolish it, and to institute new Government, laying its foundation on such principles and organizing its powers in such form, as to them shall seem most likely to effect their Safety and Happiness. Prudence, indeed, will dictate that Governments long established should not be changed for light and transient causes; and accordingly all experience hath shown, that mankind are more disposed to suffer, while evils are sufferable, than to right themselves by abolishing the forms to which they are accustomed. But when a long train of abuses and usurpations pursuing invariably the same Object evinces a design to reduce them under absolute Despotism, it is their right, it is their duty, to throw off such government, and to provide new Guards for their future security. Such has been the patient sufferance of these Colonies; and such is now the necessity which constrains them to alter their former Systems of Government. The history of the present King of Great Britain is a history of repeated injuries and usurpations, all having in direct object the establishment of an absolute Tyranny over these States. To prove this, let Facts be submitted to a candid world.

3 He has refused his Assent to Laws, the most wholesome and necessary for the public good.

4 He has forbidden his Governors to pass Laws of immediate and pressing importance, unless suspended in their operation till his Assent should be obtained; and when so suspended, he has utterly neglected to attend to them.

5 He has refused to pass other Laws for the accommodation of large districts of people, unless those people would relinquish the right of Representation in the Legislature, a right inestimable to them and formidable to tyrants only.

6 He has called together legislative bodies at places unusual, uncomfortable, and distant from the depository of their Public Records, for the sole purpose of fatiguing them into compliance with his measures.

7 He has dissolved Representative Houses repeatedly, for opposing with manly firmness his invasions on the rights of the people.

8 He has refused for a long time, after such dissolutions, to cause others to be elected; whereby the Legislative Powers, incapable of Annihilation, have returned to the People at large for their exercise; the State remaining in the mean time exposed to all the dangers of invasion from without, and convulsions within.

9 He has endeavoured to prevent the population of these States; for that purpose obstructing the Laws of Naturalization of Foreigners; refusing to pass others to encourage their migration hither, and raising the conditions of new Appropriations of Lands.

10 He has obstructed the Administration of Justice, by refusing his Assent to Laws for establishing Judiciary Powers.

11 He has made Judges dependent on his Will alone, for the tenure of their offices, and the amount and payment of their salaries.

He has erected a multitude of New Offices, and sent hither   12
swarms of Officers to harass our People, and eat out their substance.

He has kept among us, in time of peace, Standing Armies without   13
the Consent of our Legislature.

He has affected to render the Military independent of and supe-   14
rior to the Civil Power.

He has combined with others to subject us to jurisdictions foreign   15
to our constitution, and unacknowledged by our laws; giving his
Assent to their acts of pretended Legislation:

For quartering large bodies of armed troops among us: 16

For protecting them, by a mock Trial, from Punishment for any   17
Murders which they should commit on the Inhabitants of these
States:

For cutting off our Trade with all parts of the world: 18

For imposing Taxes on us without our Consent: 19

For depriving us in many cases, of the benefits of Trial by Jury: 20

For transporting us beyond Seas to be tried for pretended of-   21
fenses:

For abolishing the free System of English Laws in a Neighbour-   22
ing Province, establishing therein an Arbitrary government, and
enlarging its boundaries so as to render it at once an example and
fit instrument for introducing the same absolute rule into these
Colonies:

For taking away our Charters, abolishing our most valuable   23
Laws, and altering fundamentally the Forms of our Governments:

For suspending our own Legislatures, and declaring themselves   24
invested with Power to legislate for us in all cases whatsoever.

He has abdicated Government here, by declaring us out of his   25
Protection and waging War against us.

He has plundered our seas, ravaged our Coasts, burnt our towns   26
and destroyed the Lives of our people.

He is at this time transporting large Armies of foreign Mercenar-   27
ies to compleat works of death, desolation and tyranny, already
begun with circumstances of Cruelty & perfidy scarcely paralleled
in the most barbarous ages, and totally unworthy the Head of a
civilized nation.

He has constrained our fellow Citizens taken Captive on the high   28
Seas to bear Arms against their Country, to become the executioners
of their friends and Brethren, or to fall themselves by their Hands.

29   He has excited domestic insurrections amongst us, and has endeavoured to bring on the inhabitants of our frontiers, the merciless Indian Savages, whose known rule of warfare, is an undistinguished destruction of all ages, sexes and conditions.

30   In every stage of these Oppressions We Have Petitioned for Redress in the most humble terms: Our repeated petitions have been answered only by repeated injury. A Prince, whose character is thus marked by every act which may define a Tyrant, is unfit to be the ruler of a free People.

31   Nor have We been wanting in attention to our British brethren. We have warned them from time to time of attempts by their legislature to extend an unwarrantable jurisdiction over us. We have reminded them of the circumstances of our emigration and settlement here. We have appealed to their native justice and magnanimity and we have conjured them by the ties of our common kindred to disavow these usurpations, which would inevitably interrupt our connections and correspondence. They too have been deaf to the voice of justice and of consanguinity. We must, therefore acquiesce in the necessity, which denounces our Separation, and hold them, as we hold the rest of mankind, Enemies in War, in Peace Friends.

32   We, therefore, the Representatives of the United States of America, in General Congress, Assembled, appealing to the Supreme Judge of the world for the rectitude of our intentions, do, in the Name, and by Authority of the good People of these Colonies, solemnly publish and declare, That these United Colonies are, and of Right ought to be Free and Independent States; that they are Absolved from all Allegiance to the British Crown, and that all political connection between them and the State of Great Britain, is and ought to be totally dissolved; and that as Free and Independent States, they have full power to levy War, conclude Peace, contract Alliances, establish Commerce, and to do all other Acts and Things which Independent States may of right do. And for the support of this Declaration, with a firm reliance on the protection of Divine Providence, we mutually pledge to each other our lives, our Fortunes and our sacred Honor.

STUDY QUESTIONS

1. The core of Jefferson's argument is in paragraph 2. Diagram the argument.
2. In paragraphs 3 through 31, Jefferson lists over two dozen complaints against the king's governance. Which of the complaints support Jefferson's claim, in paragraph 2, that the king's goal is "to reduce them under an absolute Despotism"?
3. In paragraph 2, what does Jefferson consider to be the proper ends of government? Of the complaints he lists, which support that claim that the king's governance has been "destructive of these ends"?
4. Notice Jefferson's use of emotionally charged language. Does it supplement or replace argument?
5. How does Jefferson anticipate and respond to the following objection: "Perhaps the King has not protected rights as well as he should. Even so, isn't rebellion too extreme a response? Wouldn't calls for reform be more appropriate?"
6. In what way does Jefferson think "that all men are created equal"? What account of *equality* does the rest of his argument seem to require?

Reference Chapters: 11, 15

HOSEA L. MARTIN

A Few Kind Words for Affirmative Action

Hosea L. Martin is vice-president of the United Way San Francisco office. In the following selection, originally publish- ed in The Wall Street Journal, *Martin defends the practice of using race as a criterion for hiring in some cases.*

1   What with all the debate about the current versions of the Civil Rights bill, I feel it's time for me to raise my voice. I'm for affirm- ative action. I can make the argument on economic grounds—the disproportionate number of blacks out of work in this country should be enough evidence that the policy isn't taking jobs away from whites.

2   But there's a second reason for my bias. Except for a sweaty warehousing job that I was forced to take when laid off in 1984, all the jobs I've had since graduating from college in 1960 were because of affirmative action. In most cases, I was one of only a handful of black managers or professionals in an organization, and a few times I found myself to be the only one in a department. I never got around to feeling lonely, because I was too busy being grateful for being on the payroll.

3   Nor did I have gnawing doubts about my qualifications for the jobs I held. I know that it's currently popular to believe that I always sat silently in the darkest corner of the conference room, ashamed that I was hired for "political reasons," but that wasn't the way it was for me.

[Hosea L. Martin, "A Few Kind Words for Affirmative Action." *The Wall Street Journal,* April 25, 1991, op-ed page.]

The truth is, I sat as close to the boss as possible and pontificated 4
as much as anyone else at the table. I realized that somewhere there
was someone who could do my job better than I could, but I also
knew that every person in the room would have to say the same
thing if he or she were strictly honest. Every single one of us—black
and white, male and female—had been hired for reasons beyond
our being able to do the job.

That's the case with just about every job in this country. There's 5
a lot of hoopla about the U.S. being a meritocracy, but even a casual
examination of performances tells us that this is a myth.

I know that this is hard to accept. Quite often the reaction to 6
hearing it is similar to that of a swaggering gunfighter of the Old
West. "Somebody's faster'n me? You're loco!" But somewhere there
was someone quicker on the draw. Ask any ghost from Boot Hill.

It's the same in today's job market. Every person wants to believe 7
that he or she was selected over hundreds of other applicants because
he or she was the "fastest gun." 'Tain't so. There probably were
dozens of faster guns in that stack of resumes on the personnel
director's desk, but you, if you were the lucky person who got the
job, were judged to be the "best fit."

Being the best fit for a job entails more than having the best set 8
of credentials (e.g., good grades, high test scores, impressive record).
Usually, the people who apply for a particular job have met all the
criteria—education, experience, career expectations, etc.—that the
ad asked for, and you can be sure that a low-level clerk has been
instructed to screen out those applications that fail to clear this
initial barrier. What winds up on the personnel director's desk is a
stack of resumes that are astonishingly uniform. A black Stanford
MBA may have survived the cut, but you can bet that a black
high-school dropout didn't.

The task of the personnel director (or whoever does the hiring) 9
would be easy if all he or she had to do was pick the person who
could "do the job"—chances are, anyone in the stack could—but
there are other considerations that have to be made, and these
considerations aren't always limited to race and sex. Satisfying these
considerations means responding to pressures outside as well as
inside the organization.

If that Stanford MBA is selected, and finds himself the only black 10
person in his department, he shouldn't feel deflated when he learns
that he owes his good fortune to affirmative action. Eventually he'll

find out that just about everybody he's working with got special consideration for one reason or another. Some had connections who were able to get them interviewed and hired; others had attended the "right" schools; still others had been hired because they were from a particular part of the country, or were members of a particular class, religion, nationality or fraternity. (I'll never forget when, as an Army clerk, I had to type dozens of application letters for a white officer who was being discharged; he never brought up his college academic record—which I knew was horrible—but in every single letter he mentioned that he was a member of Phi-something fraternity. He got a great job.)

11   Seldom will you find a person who got a job because he was the very best at doing it. No? Come on, are you trying to tell me that Dan Quayle was the best that George Bush could find?

12   Sure, there are some exceptionally capable people, and I've observed that they don't gripe about affirmative action; they know that, sooner or later, talent will out. The run-of-the-mill plodders are the ones who complain that affirmative action is blocking their career path; take away this excuse and they'd probably blame their lack of progress on sunspots. For them, it's always something.

13   Affirmative action is needed in education as well as in the workplace. Those who criticize affirmative action in their campaign against "political correctness" are wrong. Without the policy there would be a sharp drop in the already-small number of black professionals that colleges produce to serve inner-city communities and small towns. The affirmative action programs that raised the share of minorities in medical school to 10% in 1980 from about 5% in the 1960s produced tremendous benefits to society.

14   As for me, each morning I go to work with pride and confidence. I know I can do the job. I also know that I'm a beneficiary of a law—good ol' affirmative action—that does not require an organization to hire a person who clearly doesn't have the education, credentials or skills that a job demands. Any organization that has done this is guilty of ignoring qualified minority people, and of cynically setting someone up for failure. Such an organization began with the premise that no minority person could do the job, and set out to prove it.

15   One final question: If it's true that affirmative action is cramming the offices and board rooms of corporate America with blacks, why

do I see so few of them when I'm walking around San Francisco's financial district on my lunch hour?

STUDY QUESTIONS

1. What things count as job qualifications, according to Martin?
2. Other than qualifications, what are the factors Martin says are used to judge whether someone is the "best fit" for a job?
3. Martin rejects the idea that the United States is a "meritocracy," that is, that hiring is based on qualifications. Do you think it follows from this that hiring *ought* not to be done on grounds of qualifications?
4. How many different benefits does Martin argue that affirmative action programs lead to?
5. In paragraph 12, Martin responds to the objection that affirmative action programs unfairly harm some people. On what grounds does he set aside this objection?
6. Identify the generalizations Martin uses. What role do they play in the case he is making? What is the strength of the evidence he offers for them?

Reference Chapters: 7, 15

LISA H. NEWTON

Bakke and Davis: Justice, American Style

*Lisa H. Newton was born in New Jersey in 1939. She received
her Ph.D. in Philosophy from Columbia University, and is a
professor of philosophy at Fairfield University in Connecticut.
In the following essay, Newton discusses the famous* Bakke v.
Regents of the University of California *(1978) case, in which
Bakke, a white male, sued the University of California for reverse discrimination on the grounds that he had been denied
admission to its medical school at Davis even though his
grades and test scores were significantly higher than those of
several minority students who were admitted.*

1   The use of the special minority quota or "goal" to achieve a
desirable racial mix in certain professions might appear to be an
attractive solution to the problem of justice posed by generations of
racial discrimination. Ultimately, however, the quota solution fails.
It puts an intolerable burden of injustice on a system strained by too
much of that in the past, and prolongs the terrible stereotypes of
inferiority into the indefinite future. It is a serious error to urge this
course on the American people.

2   The quota system, as employed by the University of California's
medical school at Davis or any similar institution, is unjust, for all
the same reasons that the discrimination it attempts to reverse is

[Lisa H. Newton, "Bakke and Davis: Justice, American Style." *National Forum
(The Phi Beta Kappa Journal)* LVIII, no. 1 (Winter 1978), pp. 22–23, legal
references omitted.]

unjust. It diminishes the opportunities of some candidates for a social purpose that has nothing to do with them, to make "reparation" for acts they never committed. And "they" are no homogeneous "majority": as Swedish-Americans, Irish-Americans, Americans of Polish or Jewish or Italian descent, they can claim a past history of the same irrational discrimination, poverty and cultural deprivation that now plagues Blacks and Spanish-speaking individuals. In simple justice, all applicants (except, of course, the minority of WASPs!) should have access to a "track" specially constructed for their group, if any do. And none should. The salvation of every minority in America has been strict justice, the merit system strictly applied; the Davis quota system is nothing but a suspension of justice in favor of the most recent minorities, and is flatly unfair to all the others.

The quota system is generally defended by suggesting that a little bit of injustice is far outweighted by the great social good which will follow from it; the argument envisions a fully integrated society where all discrimination will be abolished. Such a result hardly seems likely. Much more likely, if ethnic quotas are legitimated by the Court in the Bakke Case, all the other ethnic minorities will promptly organize to secure special tracks of their own, including minorities which have never previously organized at all. In these days, the advantage of a medical education is sufficiently attractive to make the effort worthwhile. As elsewhere, grave political penalties will be inflicted on legislatures and institutions that attempt to ignore these interest groups. I give Davis, and every other desirable school in the country, one decade from a Supreme Court decision favorable to quotas, to collapse under the sheer administrative weight of the hundreds of special admissions tracks and quotas it will have to maintain.

But the worst effect of the quota system is on the minorities supposedly favored by it. In the past, Blacks were socially stereotyped as less intelligent than whites because disproportionately few Blacks could get into medical school; the stereotype was the result of the very racial discrimination that it attempted to justify. Under any minority quota system, ironically, that stereotype would be tragically reinforced. From the day the Court blesses the two-track system of admissions at Davis, the word is out that Black physicians, or those of Spanish or Asian derivation, are less qualified, just a little less qualified, than their "White-Anglo" counterparts, for they did not have to meet as strict a test for admission to medical school. And that judgment will apply, as the quota applies, on the basis of race alone, for we will have no way of knowing which Blacks, Spanish or Asians were admitted in a medical school's regular competition and which were admitted on the "special minority" track. The opportunity to bury their unfavorable ethnic stereotypes by clean and public success in strictly fair competition, an opportunity that our older ethnic groups seized enthusiastically, will be denied to these "special minorities" for yet another century.

5   In short, there are no gains, for American society or for groups previously disadvantaged by it, in quota systems that attempt reparation by reverse discrimination. The larger moral question of whether we should set aside strict justice for some larger social gain, does not have to be taken up in a case like this one, where procedural injustice produces only substantive harm for all concerned. Blacks, Hispanic and other minority groups which are presently economically disadvantaged will see real progress when, and only when, the American economy expands to make room for more higher status employment for all groups. The economy is not improved in the least by special tracks and quotas for special groups; on the contrary, it is burdened by the enormous weight of the nonproductive administrative procedure required to implement them. No social purpose will be served, and no justice done, by the establishment of such procedural monsters; we should hope that the Supreme Court will see its way clear to abolishing them once and for all.

STUDY QUESTIONS

1. In paragraph 2, Newton argues that minority quotas are unjust. In paragraph 4, she argues that quotas reinforce racial stereotypes. How many distinct arguments does she raise against quotas?

2. "Justice" is a key concept in paragraphs 1 and 2 of Newton's argument. What does she mean by justice? For instance, what is "the problem of justice posed by generations of racial

discrimination"? And what is the "strict justice" that has been "the salvation of every minority in America"?

3. In paragraph 3, what does Newton state to be the position of those in favor of a quota system?

4. How do you think Newton would respond to the following statement: "Perhaps the use of quota systems will reinforce negative stereotypes of minorities. But perhaps the affected minorities would prefer the benefits of preferential policies even with that problem"?

Reference Chapters: 3, 5

PART EIGHT

On Freedom of Speech, Economic Freedom, and Government Regulation

JUSTICE LOUIS BRANDEIS

New State Ice Co. v. *Liebmann* [The Government May Prohibit Competition]

Louis D. Brandeis (1856–1941) was born in Louisville, Kentucky, attended Harvard Law School, and eventually became a U.S. Supreme Court Justice. The following case deals with the issue of whether the State can legitimately prohibit competition among private enterprises. The majority of the Court agreed with the argument for the defendant, Mr. Liebmann, that since Mr. Liebmann's ice company was a private business, the State could not legitimately prohibit him from competing with New State Ice Company. This decision was based on the Fourteenth Amendment to the Federal Constitution:

[Justice Brandeis, dissenting opinion, *New State Ice Co.* v. *Liebmann*, 463 U.S. 262 (1931), pp. 262–311, all references omitted.]

No State shall make or enforce any law which shall abridge the privileges or immunities of citizens of the United States; nor shall any State deprive any person of life, liberty, or property, without due process of law; nor deny to any person within its jurisdiction the equal protection of the laws.

The reasons for the majority's decision are summarized first; Justice Brandeis's dissenting opinion follows.

1   Argued February 19, 1932.—Decided March 21, 1932.

      1. The business of manufacturing ice and selling it is essentially a private business and not so affected with a public interest that a legislature may constitutionally limit the number of those who may engage in it, in order to control competition.

      2. An Oklahoma statute, declaring that the manufacture, sale and distribution of ice is a public business, forbids anyone to engage in it without first having procured a license from a state commission; no license is to issue without proof of necessity for the manufacture, sale or distribution of ice in the community or place to which the application relates, and if the facilities already existing and licensed at such place are sufficient to meet the public needs therein, the commission may deny the application. *Held* repugnant to the due process clause of the Fourteenth Amendment.

      3. A state law infringing the liberty guaranteed to individuals by the Constitution can not be upheld upon the ground that the State is conducting a legislative experiment.

Affirmed.

Mr. Justice Brandeis, dissenting.

2   Chapter 147 of the Session Laws of Oklahoma, 1925, declares that the manufacture of ice for sale and distribution is "a public business"; confers upon the Corporation Commission in respect to it the powers of regulation customarily exercised over public utilities; and provides specifically for securing adequate service. The statute

makes it a misdemeanor to engage in the business without a license from the Commission; directs that the license shall not issue except pursuant to a prescribed written application, after a formal hearing upon adequate notice both to the community to be served and to the general public, and a showing upon competent evidence, of the necessity "at the place desired;" and it provides that the application may be denied, among other grounds, if "the facts proved at said hearing disclose that the facilities for the manufacture, sale and distribution of ice by some person, firm or corporation already licensed by said Commission at said point, community or place are sufficient to meet the public needs therein."

Under a license, so granted, the New State Ice Company is, and for some years has been, engaged in the manufacture, sale and distribution of ice at Oklahoma City, and has invested in that business $500,000. While it was so engaged, Liebmann, without having obtained or applied for a license, purchased a parcel of land in that city and commenced the construction thereon of an ice plant for the purpose of entering the business in competition with the plaintiff. To enjoin him from doing so this suit was brought by the Ice Company. Compare *Frost* v. *Corporation Commission.* Liebmann contends that the manufacture of ice for sale and distribution is not a public business; that it is a private business and, indeed, a common calling; that the right to engage in a common calling is one of the fundamental liberties guaranteed by the due process clause; and that to make his right to engage in that calling dependent upon a finding of public necessity deprives him of liberty and property in violation of the Fourteenth Amendment. Upon full hearing the District Court sustained that contention and dismissed the bill. Its decree was affirmed by the Circuit Court of Appeals. The case is here on appeal. In my opinion, the judgment should be reversed.

First. The Oklahoma statute makes entry into the business of manufacturing ice for sale and distribution dependent, in effect, upon a certificate of public convenience and necessity. Such a certificate was unknown to the common law. It is a creature of the machine age, in which plants have displaced tools and businesses are substituted for trades. The purpose of requiring it is to promote the public interest by preventing waste. Particularly in those businesses in which interest and depreciation charges on plant constitute a large element in the cost of production, experience has taught that the financial burdens incident to unnecessary duplication of facilities are likely to bring high rates and poor service. There, cost is usually dependent, among other things, upon volume; and division of possible patronage among competing concerns may so raise the unit cost of operation as to make it impossible to provide adequate service at reasonable rates. The introduction in the United States of the certificate of public convenience and necessity marked the growing conviction that under certain circumstances free competition might be harmful to the community and that, when it was so, absolute freedom to enter the business of one's choice should be denied.

5   Long before the enactment of the Oklahoma statute here challenged a like requirement had become common in the United States in some lines of business. The certificate was required first for railroads; then for street railways; then for other public utilities whose operation is dependent upon the grant of some special privilege.

Latterly, the requirement has been widely extended to common carriers by motor vehicle which use the highways, but which, unlike street railways and electric light companies, are not dependent upon the grant of any special privilege. In Oklahoma the certificate was required, as early as 1915, for cotton gins—a business then declared a public one, and, like the business of manufacturing ice, conducted wholly upon private property. See *Frost* v. *Corporation Commission.* As applied to public utilities, the validity under the Fourteenth Amendment of the requirement of the certificate has never been successfully questioned.

6   *Second.* Oklahoma declared the business of manufacturing ice for sale and distribution a "public business"; that is, a public utility. So far as appears, it was the first State to do so. Of course, a legislature cannot by mere legislative fiat convert a business into a public utility. *Producers Transportation Co.* v. *Railroad Commission.* But the conception of a public utility is not static. The welfare of the community may require that the business of supplying ice be made a public utility, as well as the business of supplying water or any other necessary commodity or service. If the business is, or can be made, a public utility, it must be possible to make the issue of a certificate a prerequisite to engaging in it.

7   Whether the local conditions are such as to justify converting a

private business into a public one is a matter primarily for the determination of the state legislature. Its determination is subject to judicial review; but the usual presumption of validity attends the enactment. The action of the State must be held valid unless clearly arbitrary, capricious or unreasonable. "The legislature being familiar with local conditions is, primarily, the judge of the necessity of such enactments. The mere fact that a court may differ with the legislature in its views of public policy, or that judges may hold views inconsistent with the propriety of the legislation in question, affords no ground for judicial interference. . . ." *McLean* v. *Arkansas.* Whether the grievances are real or fancied, whether the remedies are wise or foolish, are not matters about which the Court may concern itself. "Our present duty is to pass upon the statute before us, and if it has been enacted upon a belief of evils that is not arbitrary we cannot measure their extent against the estimate of the legislature." *Tanner* v. *Little.* A decision that the legislature's belief of evils was arbitrary, capricious and unreasonable may not be made without enquiry into the facts with reference to which it acted.

Third. Liebmann challenges the statute—not an order of the Corporation Commission. If he had applied for a license and been denied one, we should have been obliged to enquire whether the evidence introduced before the Commission justified it in refusing permission to establish an additional ice plant in Oklahoma City. As he did not apply but challenges the statute itself, our enquiry is of an entirely different nature. Liebmann rests his defense upon the broad claim that the Federal Constitution gives him the right to enter the business of manufacturing ice for sale even if his doing so be found by the properly constituted authority to be inconsistent with the public welfare. He claims that, whatever the local conditions may demand, to confer upon the Commission power to deny that right is an unreasonable, arbitrary and capricious restraint upon his liberty.

The function of the Court is primarily to determine whether the conditions in Oklahoma are such that the legislature could not reasonably conclude (1) that the public welfare required treating the manufacture of ice for sale and distribution as a "public business"; and (2) that in order to ensure to the inhabitants of some communities an adequate supply of ice at reasonable rates it was necessary to give the Commission power to exclude the establishment of an additional ice plant in places where the community was already well served. Unless the Court can say that the Federal Constitution confers an absolute right to engage anywhere in the business of manufacturing ice for sale, it cannot properly decide that the legislators acted unreasonably without first ascertaining what was the experience of Oklahoma in respect to the ice business. The relevant facts appear, in part, of record. Others are matters of common knowledge to those familiar with the ice business. Compare *Muller* v. *Oregon.* They show the actual conditions, or the beliefs, on which the legislators acted. In considering these matters we do not, in a strict sense, take judicial notice of them as embodying statements of uncontrovertible facts. Our function is only to determine the reasonableness of the legislature's belief in the existence of evils and in the effectiveness of the remedy provided. In performing this function we have no occasion to consider whether all the statements of fact which may be the basis of the prevailing belief are well-founded; and we have, of course, no right to weigh conflicting evidence.

10   (A) In Oklahoma a regular supply of ice may reasonably be considered a necessary of life, comparable to that of water, gas and electricity. The climate, which heightens the need of ice for comfortable and wholesome living, precludes resort to the natural product. There, as elsewhere, the development of the manufactured ice industry in recent years has been attended by deep-seated alterations in the economic structure and by radical changes in habits of popular thought and living. Ice has come to be regarded as a household necessity, indispensable to the preservation of food and so to economical household management and the maintenance of health. Its commercial uses are extensive. In urban communities, they absorb a large proportion of the total amount of ice manufactured for sale. The transportation, storage and distribution of a great part of the nation's food supply is dependent upon a continuous, and dependable supply of ice. It appears from the record that in certain parts of Oklahoma a large trade in dairy and other products has been built up as a result of rulings of the Corporation Commission under the Act of 1925, compelling licensed manufacturers to serve agricultural communities; and that this trade would be destroyed if the supply of ice were withdrawn. We cannot say that the legislature of Oklahoma acted arbitrarily in declaring that ice is an article of

primary necessity, in industry and agriculture as well as in the household, partaking of the fundamental character of electricity, gas, water, transportation and communication.

Nor can the Court properly take judicial notice that, in Oklahoma, the means of manufacturing ice for private use are within the reach of all persons who are dependent upon it. Certainly it has not been so. In 1925 domestic mechanical refrigeration had scarcely emerged from the experimental stage. Since that time, the production and consumption of ice manufactured for sale, far from diminishing, has steadily increased. In Oklahoma the mechanical household refrigerator is still an article of relative luxury. Legislation essential to the protection of individuals of limited or no means is not invalidated by the circumstance that other individuals are financially able to protect themselves. The businesses of power companies and of common carriers by street railway, steam railroad or motor vehicle fall within the field of public control, although it is possible, for a relatively modest outlay, to install individual power plants, or to purchase motor vehicles for private carriage of passengers or goods. The question whether in Oklahoma the means of securing refrigeration otherwise than by ice manufactured for sale and distribution has become so general as to destroy popular dependence upon ice plants is one peculiarly appropriate for the determination of its legislature and peculiarly inappropriate for determination by this Court, which cannot have knowledge of all the relevant facts.

The business of supplying ice is not only a necessity, like that of supplying food or clothing or shelter, but the legislature could also consider that it is one which lends itself peculiarly to monopoly. Characteristically the business is conducted in local plants with a market narrowly limited in area, and this for the reason that ice manufactured at a distance cannot effectively compete with a plant on the ground. In small towns and rural communities the duplication of plants, and in larger communities the duplication of delivery service, is wasteful and ultimately burdensome to consumers. At the same time the relative ease and cheapness with which an ice plant may be constructed exposes the industry to destructive and frequently ruinous competition. Competition in the industry tends to be destructive because ice plants have a determinate capacity, and inflexible fixed charges and operating costs, and because in a market

11

12

of limited area the volume of sales is not readily expanded. Thus, the erection of a new plant in a locality already adequately served often causes managers to go to extremes in cutting prices in order to secure business. Trade journals and reports of association meetings of ice manufacturers bear ample witness to the hostility of the industry to such competition, and to its unremitting efforts, through trade associations, informal agreements, combination of delivery systems, and in particular through the consolidation of plants, to protect markets and prices against competition of any character.

13   That these forces were operative in Oklahoma prior to the passage of the Act under review, is apparent from the record. Thus, it was testified that in only six or seven localities in the State containing, in the aggregate, not more than 235,000 of the total population of approximately 2,000,000, was there "a semblance of competition"; and that even in those localities the prices of ice were ordinarily uniform. The balance of the population was, and still is, served by companies enjoying complete monopoly. Compare *Munn* v. *Illinois; Sinking Fund Cases; Wabash, St. L. & P. Ry. Co.* v. *Illinois; Spring Valley Water Works* v. *Schottler; Budd* v. *New York; Wolff Co.* v. *Industrial Court.* Where there was competition, it often resulted to the disadvantage rather than the advantage of the public, both in respect to prices and to service. Some communities were without ice altogether, and the State was without means of assuring their supply. There is abundant evidence of widespread dissatisfaction with ice service prior to the Act of 1925, and of material improvement in the situation subsequently. It is stipulated in the record that the ice industry as a whole in Oklahoma has acquiesced in and accepted the Act and the status which it creates.

14   (B) The statute under review rests not only upon the facts just detailed but upon a long period of experience in more limited regulation dating back to the first year of Oklahoma's statehood. For 17 years prior to the passage of the Act of 1925, the Corporation Commission under §13 of the Act of June 10, 1908, had exercised jurisdiction over the rates, practices and service of ice plants, its action in each case, however, being predicated upon a finding that the company complained of enjoyed a "virtual monopoly" of the ice business in the community which it served. The jurisdiction thus exercised was upheld by the Supreme Court of the State in *Oklahoma Light & Power Co.* v. *Corporation Commission.* The court said,

at p. 24: "The manufacture, sale, and distribution of ice in many respects closely resemble the sale and distribution of gas as fuel, or electric current, and in many communities the same company that manufactures, sells, and distributes electric current is the only concern that manufactures, sells, and distributes ice, and by reason of the nature and extent of the ice business it is impracticable in that community to interest any other concern in such business. In this situation, the distributor of such a necessity as ice should not be permitted by reason of the impracticability of any one else engaging in the business to charge unreasonable prices, and if such an abuse is persisted in, the regulatory power of the State should be invoked to protect the public." See also *Consumers Light & Power Co.* v. *Phipps.*

By formal orders, the Commission repeatedly fixed or approved   15
prices to be charged in particular communities; required ice to be sold without discrimination and to be distributed as equitably as possible to the extent of the capacity of the plant; forbade short weights and ordered scales to be carried on delivery wagons and ice to be weighed upon the customer's request; and undertook to compel sanitary practices in the manufacture of ice and courteous service of patrons. Many of these regulations, other than those fixing prices, were embodied in a general order to all ice companies, issued July 15, 1921, and are still in effect. Informally, the Commission adjusted a much greater volume of complaints of a similar nature. It appears from the record that for some years prior to the Act of 1925 one day of each week was reserved by the Commission to hear complaints relative to the ice business.

As early as 1911, the Commission in its annual report to the   16
Governor, had recommended legislation more clearly delineating its powers in this field:

> There should be a law passed putting the regulation of ice plants   17
> under the jurisdiction of the Commission. The Commission is now
> assuming this jurisdiction under an Act passed by the Legislature
> known as the anti-trust law. A specific law upon this subject would
> obviate any question of jurisdiction.

This recommendation was several times repeated, in terms revealing the extent and character of public complaint against the practices of ice companies.

18   The enactment of the so-called Ice Act in 1925 enlarged the existing jurisdiction of the Corporation Commission by removing the requirement of a finding of virtual monopoly in each particular case, compare *Budd* v. *New York* with *Brass* v. *Stoeser;* by conferring the same authority to compel adequate service as in the case of other public utilities; and by committing to the Commission the function of issuing licenses equivalent to a certificate of public convenience and necessity. With the exception of the granting and denying of such licenses and the exertion of wider control over service, the regulatory activity of the Commission in respect to ice plants has not changed in character since 1925. It appears to have diminished somewhat in volume.

19   In 1916, the Commission urged, in its report to the Governor, that all public utilities under its jurisdiction be required to secure from the Commission "what is known as a 'certificate of public convenience and necessity' before the duplication of facilities.

20   "This would prevent ruinous competition resulting in the driving out of business of small though competent public service utilities by more powerful corporations, and often consequent demoralization of service, or the requiring of the public to patronize two utilities in a community where one would be adequate."

21   Up to that time a certificate of public convenience and necessity to engage in the business had been applied only to cotton gins. In 1917 a certificate from the Commission was declared prerequisite to the construction of new telephone or telegraph lines. In 1923 it was required for the operation of motor carriers. In 1925, the year in which the Ice Act was passed, the requirement was extended also to power, heat, light, gas, electric or water companies proposing to do business in any locality already possessing one such utility.

22   *Fourth.* Can it be said in the light of these facts that it was not an appropriate exercise of legislative discretion to authorize the Commission to deny a license to enter the business in localities where necessity for another plant did not exist? The need of some remedy for the evil of destructive competition, where competition existed, had been and was widely felt. Where competition did not exist, the propriety of public regulation had been proven. Many communities were not supplied with ice at all. The particular remedy adopted was not enacted hastily. The statute was based upon a long-established state policy recognizing the public importance of the ice

business, and upon 17 years' legislative and administrative experience in the regulation of it. The advisability of treating the ice business as a public utility and of applying to it the certificate of convenience and necessity had been under consideration for many years. Similar legislation had been enacted in Oklahoma under similar circumstances with respect to other public services. The measure bore a substantial relation to the evils found to exist. Under these circumstances, to hold the Act void as being unreasonable, would, in my opinion involve the exercise not of the function of judicial review, but the function of a super-legislature. If the Act is to be stricken down, it must be on the ground that the Federal Constitution guarantees to the individual the absolute right to enter the ice business, however detrimental the exercise of that right may be to the public welfare. Such, indeed, appears to be the contention made

Fifth. The claim is that manufacturing ice for sale and distribution is a business inherently private, and, in effect, that no state of facts can justify denial of the right to engage in it. To supply one's self with water, electricity, gas, ice or any other article, is inherently a matter of private concern. So also may be the business of supplying the same articles to others for compensation. But the business of supplying to others, for compensation, any article or service whatsoever may become a matter of public concern. Whether it is, or is not, depends upon the conditions existing in the community affected. If it is a matter of public concern, it may be regulated, whatever the business. The public's concern may be limited to a single feature of the business, so that the needed protection can be secured by a relatively slight degree of regulation. Such is the concern over possible incompetence, which dictates the licensing of dentists, *Dent* v. *West Virginia, Douglas* v. *Noble,* or the concern over possible dishonesty, which led to the licensing of auctioneers or hawkers, *Baccus* v. *Louisiana.* On the other hand, the public's concern about a particular business may be so pervasive and varied as to require constant detailed supervision and a very high degree of regulation. Where this is true, it is common to speak of the business as being a "public" one, although it is privately owned. It is to such businesses that the designation "public utility" is commonly applied; or they are spoken of as "affected with a public interest." *German Alliance Ins. Co.* v. *Lewis.*

23

24   A regulation valid for one kind of business may, of course, be invalid for another; since the reasonableness of every regulation is dependent upon the relevant facts. But so far as concerns the power to regulate, there is no difference in essence, between a business called private and one called a public utility or said to be "affected with a public interest." Whatever the nature of the business, whatever the scope or character of the regulation applied, the source of the power invoked is the same. And likewise the constitutional limitation upon that power. The source is the police power. The limitation is that set by the due process clause, which, as construed, requires that the regulation shall be not unreasonable, arbitrary or capricious; and that the means of regulation selected shall have a real or substantial relation to the object sought to be obtained. The notion of a distinct category of business "affected with a public interest," employing property "devoted to a public use," rests upon historical error. The consequences which it is sought to draw from those phrases are belied by the meaning in which they were first used centuries ago, and by the decision of this Court, in *Munn* v. *Illinois,* which first introduced them into the law of the Constitution. In my opinion, the true principle is that the State's power extends to every regulation of any business reasonably required and appropriate for the public protection. I find in the due process clause no other limitation upon the character or the scope of regulation permissible.

25   *Sixth.* It is urged specifically that manufacturing ice for sale and distribution is a common calling; and that the right to engage in a common calling is one of the fundamental liberties guaranteed by the due process clause. To think of the ice-manufacturing business as a common calling is difficult; so recent is it in origin and so peculiar in character. Moreover, the Constitution does not require that every calling which has been common shall ever remain so. The liberty to engage in a common calling, like other liberties, may be limited in the exercise of the police power. The slaughtering of cattle had been a common calling in New Orleans before the monopoly sustained in *Slaughter-House Cases,* was created by the legislature. Prior to the Eighteenth Amendment selling liquor was a common calling, but this Court held it to be consistent with the due process clause for a State to abolish the calling, *Bartemeyer* v. *Iowa, Mugler* v. *Kansas,* or to establish a system limiting the number

of licenses, *Crowley* v. *Christensen.* Every citizen has the right to navigate a river or lake, and may even carry others thereon for hire. But the ferry privilege may be made exclusive in order that the patronage may be sufficient to justify maintaining the ferry service, *Conway* v. *Taylor's Executor.*

It is settled that the police power commonly invoked in aid of 26 health, safety and morals, extends equally to the promotion of the public welfare. The cases just cited show that, while, ordinarily, free competition in the common callings has been encouraged, the public welfare may at other times demand that monopolies be created. Upon this principle is based our whole modern practice of public utility regulation. It is no objection to the validity of the statute here assailed that it fosters monopoly. That, indeed, is its design. The certificate of public convenience and invention is a device—a recent social-economic invention—through which the monopoly is kept under effective control by vesting in a commission the power to terminate it whenever that course is required in the public interest. To grant any monopoly to any person as a favor is forbidden even if terminable. But where, as here, there is reasonable ground for the legislative conclusion that in order to secure a necessary service at reasonable rates, it may be necessary to curtail the right to enter the calling, it is, in my opinion, consistent with the due process clause to do so, whatever the nature of the business. The existence of such power in the legislature seems indispensable in our ever-changing society.

It is settled by unanimous decisions of this Court, that the due 27 process clause does not prevent a State or city from engaging in the business of supplying its inhabitants with articles in general use, when it is believed that they cannot be secured at reasonable prices from the private dealers. Thus, a city may, if the local law permits, buy and sell at retail coal and wood, *Jones* v. *Portland,* or gasoline, *Standard Oil Co.* v. *Lincoln.* And a State may, if permitted by its own Constitution, build and operate warehouses, elevators, packing-houses, flour mills or other factories, *Green* v. *Frazier.* As States may engage in a business, because it is a public purpose to assure to their inhabitants an adequate supply of necessary articles, may they not achieve this public purpose, as Oklahoma has done, by exercising the lesser power of preventing single individuals from wantonly engaging in the business and thereby making impossible a dependable private source of supply? As a State so entering upon a business may exert the taxing power all individual dealers may be driven from the calling by the unequal competition. If States are denied the power to prevent the harmful entry of a few individuals into a business, they may thus, in effect, close it altogether to private enterprise.

28   *Seventh.* The economic emergencies of the past were incidents of scarcity. In those days it was preëminently the common callings that were the subjects of regulation. The danger then threatening was excessive prices. To prevent what was deemed extortion, the English Parliament fixed the prices of commodities and of services from time to time during the four centuries preceding the Declaration of Independence. Like legislation was enacted in the Colonies; and in the States, after the Revolution. When the first due process clause was written into the Federal Constitution, the price of bread was being fixed by statute in at least two of the States, and this practice continued long thereafter. Dwelling houses when occupied by the owner are preëminently private property. From the foundation of our Government those who wished to lease residential property had been free to charge to tenants such rentals as they pleased. But for years after the World War had ended, the scarcity of dwellings in the City of New York was such that the State's legislative power was invoked to ensure reasonable rentals. The constitutionality of the statute was sustained by this Court. *Marcus Brown Holding Co.* v. *Feldman.* Similar legislation of Congress for the City of Washington was also upheld. *Block* v. *Hirsh.*

29   *Eighth.* The people of the United States are now confronted with an emergency more serious than war. Misery is wide-spread, in a time, not of scarcity, but of over-abundance. The long-continued depression has brought unprecedented unemployment, a catastrophic fall in commodity prices and a volume of economic losses which threatens our financial institutions. Some people believe that the existing conditions threaten even the stability of the capitalistic system. Economists are searching for the causes of this disorder and are reexamining the bases of our industrial structure. Business men are seeking possible remedies. Most of them realize that failure to distribute widely the profits of industry has been a prime cause of our present plight. But rightly or wrongly, many persons think that one of the major contributing causes has been unbridled competi-

tion. Increasingly, doubt is expressed whether it is economically wise, or morally right, that men should be permitted to add to the producing facilities of an industry which is already suffering from over-capacity. In justification of that doubt, men point to the excess-capacity of our productive facilities resulting from their vast expansion without corresponding increase in the consumptive capacity of the people. They assert that through improved methods of manufacture, made possible by advances in science and invention and vast accumulation of capital, our industries had become capable of producing from thirty to one hundred per cent. more than was consumed even in days of vaunted prosperity; and that the present capacity will, for a long time, exceed the needs of business. All agree that irregularity in employment—the greatest of our evils—cannot be overcome unless production and consumption are more nearly balanced. Many insist there must be some form of economic control. There are plans for proration. There are many proposals for stabilization. And some thoughtful men of wide business experience insist that all projects for stabilization and proration must prove futile unless, in some way, the equivalent of the certificate of public convenience and necessity is made a prerequisite to embarking new capital in an industry in which the capacity already exceeds the production schedules.

Whether that view is sound nobody knows. The objections to the proposal are obvious and grave. The remedy might bring evils worse than the present disease. The obstacles to success seem insuperable. The economic and social sciences are largely uncharted seas. We have been none too successful in the modest essays in economic control already entered upon. The new proposal involves a vast extension of the area of control. Merely to acquire the knowledge essential as a basis for the exercise of this multitude of judgments would be a formidable task; and each of the thousands of these judgments would call for some measure of prophecy. Even more serious are the obstacles to success inherent in the demands which execution of the project would make upon human intelligence and upon the character of men. Man is weak and his judgment is at best fallible.

Yet the advances in the exact sciences and the achievements in invention remind us that the seemingly impossible sometimes happens. There are many men now living who were in the habit of using the age-old expression: "It is as impossible as flying." The discoveries in physical science, the triumphs in invention, attest the value of the process of trial and error. In large measure, these advances have been due to experimentation. In those fields experimentation has, for two centuries, been not only free but encouraged. Some people assert that our present plight is due, in part, to the limitations set by courts upon experimentation in the fields of social and economic science; and to the discouragement to which proposals for betterment there have been subjected otherwise. There must be power in the States and the Nation to remould, through experimentation, our economic practices and institutions to meet changing social and economic needs. I cannot believe that the framers of the Fourteenth Amendment, or the States which ratified it, intended to deprive us of the power to correct the evils of technological unemployment and excess productive capacity which have attended progress in the useful arts.

32 To stay experimentation in things social and economic is a grave responsibility. Denial of the right to experiment may be fraught with serious consequences to the Nation. It is one of the happy incidents of the federal system that a single courageous State may, if its citizens choose, serve as a laboratory; and try novel social and economic experiments without risk to the rest of the country. This Court has the power to prevent an experiment. We may strike down the statute which embodies it on the ground that, in our opinion, the measure is arbitrary, capricious or unreasonable. We have power to do this, because the due process clause has been held by the Court applicable to matters of substantive law as well as to matters of procedure. But in the exercise of this high power, we must be ever on our guard, lest we erect our prejudices into legal principles. If we would guide by the light of reason, we must let our minds be bold.

STUDY QUESTIONS

 1. Brandeis's overall conclusion is that the State of Oklahoma was justified in not allowing Liebmann to manufacture ice for sale. To support this conclusion he argues three points:
  (i) That prohibiting competition in this case was to the public's welfare;

(ii) That there are precedents for states' creating monopolies as public utilities;

(iii) That the State's action does not violate the Fourteenth Amendment.

Here is a reconstruction of his argument in support of (i):

> The State's function is to promote the public welfare.
> It is to the public's welfare not to have waste, expensive products, or poor service.
> So the State should not to allow waste, expensive products, or poor service.
> Sometimes competition leads to waste, expensive products, or poor service.
> So sometimes the State should not to allow competition.
> Competition in the ice business in Oklahoma led to waste and poor service.
> Therefore, it was legitimate for the State of Oklahoma not to allow competition in the ice business.

For each of the premises in this argument, determine whether Brandeis assumes its truth or provides an argument for it. And for those premises he argues for, diagram the arguments.

2. "The public welfare" is the key phrase in Brandeis's argument. Does he offer a definition of it, or list the criteria by which we can tell what is or is not to the public's welfare?

3. Brandeis discusses the Fourteenth Amendment to the Federal Constitution in paragraphs 3, 8, 9, and 22–27. He argues against interpreting the Fourteenth Amendment as giving an individual an "absolute" right to engage in business. Write a short paragraph summarizing how Brandeis interprets the Fourteenth Amendment. Under what circumstances does he argue that the right can be limited?

4. Has Brandeis argued successfully that the State of Oklahoma neither abridged Liebmann's privileges or immunities, nor deprived him of liberty or property, nor denied him equal protection of the laws?

Reference Chapters: 3, 7, 11

HENRY HAZLITT

Who's "Protected" by Tariffs?

Henry Hazlitt (1894–1993) is best known for his clear and graceful defenses of free market economic concepts. Drawing on the work of economists Claude-Frédéric Bastiat, Philip Wicksteed, and Ludwig von Mises, Hazlitt wrote the popular Economics in One Lesson, *from which the following excerpt is taken. He also wrote a novel entitled* Time Will Run Back.

1   An American manufacturer of woolen sweaters goes to Congress or to the State Department and tells the committee or officials concerned that it would be a national disaster for them to remove or reduce the tariff on British sweaters. He now sells his sweaters for $30 each, but English manufacturers could sell their sweaters of the same quality for $25. A duty of $5, therefore, is needed to keep him in business. He is not thinking of himself, of course, but of the thousand men and women he employs, and of the people to whom their spending in turn gives employment. Throw them out of work, and you create unemployment and a fall in purchasing power, which would spread in ever-widening circles. And if he can prove that he really would be forced out of business if the tariff were removed or reduced, his argument against that action is regarded by Congress as conclusive.

2   But the fallacy comes from looking merely at this manufacturer and his employees, or merely at the American sweater industry. It comes from noticing only the results that are immediately seen, and

[Henry Hazlitt, excerpt from Chapter XI of *Economics in One Lesson.* Westport, Conn.: Arlington House, 1946; reprinted 1962, 1979; pp. 75–77.]

neglecting the results that are not seen because they are prevented from coming into existence.

The lobbyists for tariff protection are continually putting forward arguments that are not factually correct. But let us assume that the facts in this case are precisely as the sweater manufacturer has stated them. Let us assume that a tariff of $5 a sweater is necessary for him to stay in business and provide employment at sweater-making for his workers.

We have deliberately chosen the most unfavorable example of any for the removal of a tariff. We have not taken an argument for the imposition of a new tariff in order to bring a new industry into existence, but an argument for the retention of a tariff *that has already brought an industry into existence,* and cannot be repealed without hurting somebody.

The tariff is repealed; the manufacturer goes out of business; a thousand workers are laid off; the particular tradesmen whom they patronized are hurt. This is the immediate result that is seen. But there are also results which, while much more difficult to trace, are no less immediate and no less real. For now sweaters that formerly cost retail $30 apiece can be bought for $25. Consumers can now buy the same quality of sweater for less money, or a much better one for the same money. If they buy the same quality of sweater, they not only get the sweater, but they have $5 left over, which they would not have had under the previous conditions, to buy something else. With the $25 that they pay for the imported sweater they help employment—as the American manufacturer no doubt predicted—in the sweater industry in England. With the $5 left over they help employment in any number of other industries in the United States.

But the results do not end there. By buying English sweaters they furnish the English with dollars to buy American goods here. This, in fact (if I may here disregard such complications as fluctuating exchange rates, loans, credits, etc.), is the only way in which the British can eventually make use of these dollars. Because we have permitted the British to sell more to us, they are now able to buy more from us. They are, in fact, eventually *forced* to buy more from us if their dollar balances are not to remain perpetually unused. So as a result of letting in more British goods, we must export more American goods. And though fewer people are now employed in the American sweater industry, more people are employed—and much more efficiently employed—in, say, the American washing-machine or aircraft-building business. American employment on net balance has not gone down, but American and British production on net balance has gone up. Labor in each country is more fully employed in doing just those things that it does best, instead of being forced to do things that it does inefficiently or badly. Consumers in both countries are better off. They are able to buy what they want where they can get it cheapest. American consumers are better provided with sweaters, and British consumers are better provided with washing machines and aircraft.

STUDY QUESTIONS

1. In the first paragraph, Hazlitt summarizes the pro-tariff argument he is attacking. The argument is a hypothetical one:

   If the tariff of $5 is repealed, then the sweater manufacturer will be forced out of business by English competitors.
   If the sweater manufacturer is forced out of business, then a thousand men and women will be put out of work.
   If a thousand men and women are put out of work, then they have less purchasing power.
   If the thousand men and women have less purchasing power, then this fact is felt in ever-widening circles (i.e., the U.S. economy as a whole suffers).
   [Therefore, if the tariff of $5 is repealed, the U.S. economy suffers.]
   [We don't want the U.S. economy to suffer.]
   [Therefore, we should not repeal the tariff of $5.]

   Although the pro-tariff argument is valid, Hazlitt says it commits a fallacy. How, in general terms, does Hazlitt describe the fallacy he thinks is committed?

2. In paragraph 5, Hazlitt asks us to suppose that the tariff is repealed, and toward the end of paragraph 6 he reaches the conclusions that "American employment on net balance has not gone down" and that "Consumers in both countries are better off." Reconstruct the chain of reasoning Hazlitt uses to reach those conclusions.

3. One of the premises in Hazlitt's argument is that "By buying English sweaters they [American consumers] furnish the English with dollars to buy American goods here." Why couldn't the English use those dollars to buy goods from other countries? Why does Hazlitt say they are *"forced"* to buy more U.S. goods?

Reference Chapters: 10, 11

STEVEN SCHLOSSSTEIN

Professionalism Not Likely to Stop Liberal Bias of Media

Steven Schlossstein lives in Mercer County, New Jersey. He is an international strategic consultant and the author of The End of the American Century.

One of my favorite stories about the press is in the form of an anecdote showing how America's three leading newspapers might chronicle Armageddon when the end of the world finally comes.   1

"World Comes to an End," would be the lead headline in the *New York Times,* followed by a predictable subhead: "Third World Nations Hardest Hit."   2

The Washington Post, too, would lead with "World Comes to an End," but its typical subhead would read, "White House Could Have Averted Crisis, Inside Sources Say."   3

Not least, *USA Today* would run something like this: "We're   4

[Steven Schlossstein, "Professionalism Not Likely to Stop Liberal Bias of Media." *The Times* (Trenton, N.J.), February 17, 1991, op-ed page.]

Dead" with two subheads: "State-by-State Roundup, 3B; Final Final Sports, 1D."

5   While whimsical, these characterizations nonetheless contain a kernel of truth, because newspapers do, over time, assume personalities that reflect an unintentional but inherent bias.

6   In the American press, that bias most frequently tends to be liberal. When, for example, the Heritage Foundation or the American Enterprise Institute issues the results of a study or poll, the media invariably report their findings by inserting the word "conservative" as a preconditioning modifier.

7   "According to a report issued by the conservative Heritage Foundation in Washington—" was the way one typical news account recently read, leaving the reader with the distinct impression that its data was somehow biased, flawed or suspect.

8   If, however, a report is released by the Brookings Institution or the Urban Institute, two Washington think tanks known for their support of liberal policy positions, the press omits a modifier altogether, giving the reader a subconscious nudge that this data is objective, more analytical and untainted by bias.

9   In a poll conducted several years ago by Stephen Hess of the (liberal) Brookings Institution, Hess queried 1,250 reporters in Washington as follows: "Many feel that there is a bias in the Washington press corps. Do you agree?" He got 300 responses.

10   Of the respondents, 51 percent said yes and 49 percent no. Hess then asked, "In which direction is that bias?" In response, 96 percent said liberal, and only 1 percent said conservative.

11   A much larger and more comprehensive survey was conducted by the *Los Angeles Times* in 1985, consisting of 106 questions asked via half-hour telephone interviews with 2,703 news and editorial staff members of 621 newspapers of all sizes around the country.

12   These specific newspapers were selected because they had been identified by 2,993 members of the reading public as the papers they read. The results were quite revealing.

13   In the poll, 55 percent of the newspaper journalists and their editors said they were liberal, 26 percent "middle of the road," and 17 percent conservative. But only 24 percent of their readers identified themselves as liberal, 33 percent said they were moderate, and 29 percent said they were conservative.

14   The newspaper reporters and editors were not only more liberal

than their readers, but overwhelmingly white (96 percent), male (73 percent), and college-educated (88 percent).

Moreover, the journalists were much less likely than their readers   15
to consider personal or social issues such as crime, drugs, inflation or unemployment as "the biggest problem facing the country today."

In contrast, they were much more likely to indicate the "federal   16
budget deficit" or "nuclear arms control" as "the biggest problem facing the country today."

When it comes to television, there is an interesting shuttle   17
pattern that occurs between the broadcast media and the staffs of our liberal Democratic politicians in Washington. Consider:

In 1985, Dotty Lynch, Gary Hart's pollster, became political   18
editor at CBS. In 1986, when Brian Lunde, executive director of the Democratic National Committee, left that post, he was replaced by Wally Chalmers, a political researcher for CBS News and a former member of Teddy Kennedy's staff. David Burke, who was Kennedy's chief of staff from 1965 to 1971, became vice president of ABC News in 1977 and executive vice president in 1986.

In 1987, the Media Research Center identified 48 such people   19
with connections to liberal or Democratic groups who had moved from political policy jobs to the TV networks or the major print media in positions responsible for news content.

Does it matter? Do their personal liberal preferences automat-   20
ically prejudice or influence the way they report the news? They say no, of course, because their "professionalism" prevents that from happening. But one has to wonder.

But as Al Hunt observed in *The Wall Street Journal,* what if a   21
judicial nominee said he was a racist but that this personal view wouldn't adversely influence his views on civil rights?

Or what if the head of the Environmental Protection Agency said   22
that his ownership of Union Carbide stock wouldn't prejudice his view of that company's toxic waste practices?

Or what if a White House aide or congressional staff member   23
said their vacations at a particular company's resort would not be a factor in considering legislative revision of the tax code?

Would we buy those arguments? Of course not. Then should we   24
accept the statements of liberal journalists that their political views don't influence the way they report the news?

STUDY QUESTIONS

1. What conclusion does Schlossstein draw from the results of the two surveys he cites? Using the criteria for evaluating inductive and statistical arguments, determine the strength of Schlossstein's argument.
2. Schlossstein uses several analogies to support his conclusion that the liberal views of the media lead to biased reporting of news. Using the criteria for evaluating analogical arguments, determine the strength of his argument.
3. What seems to be Schlossstein's core complaint? Is it that he thinks many journalists are intentionally (or unintentionally) presenting inaccurate news? Is it that he thinks news reporting and commentary should be kept separate? Is it that he thinks that the distribution of opinions presented in journalism should match the distribution of opinions in the general population? Or is it something else?

Reference Chapters: 15, 16, 17

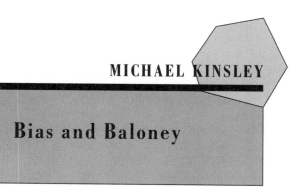

MICHAEL KINSLEY

Bias and Baloney

Michael Kinsley was born in 1951. He received his A.B and J.D. degrees from Harvard University, and is currently a sen-

[Michael Kinsley, "Bias and Baloney." *The New Republic*, December 14, 1992, pp. 6 and 45.]

ior editor at New Republic *magazine and a contributing writer for* Time.

Aha! Gotcha! That undoubtedly is the reaction of conservatives across the land to evidence that American journalists tend to have liberal political views. The evidence comes in a survey of 1,400 journalists conducted over the summer and reported in "The American Journalist in the 1990s," a study published by something called the Freedom Forum, a philanthropic spin-off of the Gannett media empire.

Few tears have been shed over the study's conclusion that the typical journalist earns less in real terms today than in the late 1960s. (Average age, 36; median 1992 income, $31,297.) Instead there has been gnashing of teeth over the news that 44.1 percent of journalists consider themselves Democrats and only 16.3 percent Republicans. The gap has grown since a similar survey in 1982 and is far larger than the gap in polls of the general population, some of which have shown the parties close to even. At a time when Republicans are toying with a variety of stab-in-the-back theories of why they lost the last election, most involving the press, such findings are red meat to right-wing conspiratorialists.

Why do journalists tend to be liberals? The extent of this tendency is exaggerated in paranoid conservative minds. And there is the countervailing reality that media owners and editorial pages tend to be conservative. Two decades after George Will was discovered sipping a cherry soda at a local drugstore, the Washington punditocracy—on the op-ed pages and T.V. yack shows—also has a definite conservative slant. Nevertheless, the general liberal inclination of journalists would be hard to deny.

Perhaps I am not the best person to try to solve the mystery of why. My own political views are more or less liberal. They were not genetically implanted, and I hold them under no form of compulsion except that of reason. It seems to me they're the sort of views a reasonable, intelligent person would hold. Since most journalists I meet are reasonable, intelligent people, the mystery to me is not why journalists tend to be liberals but why so many other reasonable, intelligent people are not. But it's not hard to come up with plausible theories to explain the gap, having to do with psychological disposition and so on.

The point is that there is no conspiracy going on here. People freely choose their politics and freely choose their careers. No one is forcing journalists to hold liberal political views, and no one is preventing or even discouraging conservatives from becoming journalists. If it just happens to work out that way, on average, so what?

6   A political preference is not itself a "bias." This basic point seems beyond the understanding of many press critics. Any journalist who has had this argument with a non-journalist knows the frustration of being accused of "bias" by someone whose political views are of such red-hot intensity that your own pale by comparison. Many who will wave this Gannett study as proof of a press liberal "bias" will refuse to accept that their conservative views make them "biased" too.

7   But should journalists be different? Some media critics, and some journalists themselves, think that the press ought to function as a sort of sacred priesthood of political celibates, purged of the ideological longings that inflame ordinary folks. The executive editor of *The Washington Post* famously goes so far as to refrain from voting. Perhaps he also succeeds in having no opinion about whom he would vote for were he not so scrupulous. If so, his self-discipline would do credit to any monk. Whether it is quite so admirable for a journalist is another question. In any event, it is surely rare. Journalists, by definition, are inquisitive people with an interest in public affairs. To expect them to form no conclusions about the people and policies they report on is absurd.

8   What do conservative media critics want? Presumably they would not favor a quota program for right-wingers, some kind of Americans with Political Disabilities Act, whereby people handicapped with conservative political opinions would get preference over better-qualified liberals for the same job. What they, and everybody else, can reasonably expect is for reporters to tell the story as straight as possible. Evidence about journalists' political preference says nothing one way or another about how they are performing that function. Most national reporters, on T.V. and in print, do it pretty well. Certainly of the two newspapers here in Washington, the one whose writers and editors make no effort to avoid slanting the news is the conservative *Washington Times*, not the supposedly liberal *Post*.

9   In Europe they do things differently. There, most newspapers are overtly ideological, and you take that into account in choosing a

paper and in how you read it. There's much to be said for this system. Opinion journalism, if it's honest, can sometimes be a more efficient way of acquiring information than an American-style objective news story (in my opinion). Intelligent judgment is less paralyzed by the need for "balance," and less space is wasted on quotations from others saying things the writer dares not say herself. There are subjects on which I'd rather read a *Wall Street Journal* editorial, undoubtedly "biased" though it may be, than a *Wall Street Journal* news article.

But in this country we cling to the belief that the news, unlike   10
editorials, should be "balanced." Thus two other silly studies, released recently, purporting to compute whether television reports and/or newspaper stories during the election campaign were pro- or anti-one candidate or another. One study, from the Joan Shorenstein Barone Center on the Press, Politics, and Public Policy at Harvard University's John F. Kennedy School of Government (an organization whose very name takes up half of any news story dedicated to reporting its findings), actually measured hundreds of news stories on a scale of one (very positive) to five (very negative).

The study, which covered February through May, found that   11
Bush's average score was 3.3—a full 0.9 more negative than Clinton's 2.4. The implicit premise of this pseudo-scientific exercise seems to be that in a perfect world every candidate would score an exact 2.5. But at a time when Bush was presiding over a stagnant economy, running an inept campaign, and being bashed from inside and outside his own party, a perfectly "balanced" press coverage would itself be evidence of bias.

The press brings these studies on itself. Not by displaying bias,   12
but by its hunger for respectability and professionalization. That's what leads to the creation of things like the Freedom Forum and the Barone Blah Blah Blah, which then need to keep themselves busy by commissioning studies and heartaching over the meaning of it all. The expanding supply of solutions creates an increasing demand for problems.

Then, too, journalists have a psychological disposition—related,   13
perhaps, to the psychological disposition that leads them to be disproportionately liberal—toward public wallowing in self-doubt and self-flagellation. Conservative press critics suffer no such malady. Or at least so far as I've noticed.

STUDY QUESTIONS

1. What is a "bias"? Why, in paragraph 6, is Kinsley concerned to distinguish a political preference from a political bias?
2. Is Kinsley arguing that the press's liberal tendencies do not lead to opinionated (or possibly biased) reporting? Or is he arguing that while liberally opinionated (or possibly biased) reporting exists, such reporting is a good thing as long as it's honest?
3. How many distinct generalizations about members of the press, liberals and conservatives, does Kinsley make in his final two paragraphs?
4. In paragraph 12, is Kinsley suggesting that journalists should not bother about respectability and professionalization? If so, why? If not, what is his purpose in raising this point?
5. Does Kinsley address all of the concerns raised in Steven Schlossstein's piece on the liberal bias of the media? For example, does he address the worry that some journalists present biased reporting under the cover of objectivity?

Reference Chapters: 3, 5

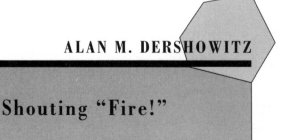

ALAN M. DERSHOWITZ

Shouting "Fire!"

Alan M. Dershowitz is an attorney, professor of law at Harvard University, and author of a weekly syndicated column as

[Alan M. Dershowitz, "Shouting 'Fire!'" *Atlantic Monthly* (January 1989), pp. 72–74.]

well as several books, the most recent of which is Chutzpah
(1991).

When the Reverend Jerry Falwell learned that the Supreme Court   1
had reversed his $200,000 judgment against *Hustler* magazine for
the emotional distress that he had suffered from an outrageous
parody, his response was typical of those who seek to censor speech:
"Just as no person may scream 'Fire!' in a crowded theater when
there is no fire, and find cover under the First Amendment, likewise,
no sleazy merchant like Larry Flynt should be able to use the First
Amendment as an excuse for maliciously and dishonestly attacking
public figures, as he has so often done."

Justice Oliver Wendell Holmes's classic example of unprotected   2
speech—falsely shouting "Fire!" in a crowded theater—has been
invoked so often, by so many people, in such diverse contexts, that
it has become part of our national folk language. It has even
appeared—most appropriately—in the theater: in Tom Stoppard's
play *Rosencrantz and Guildenstern Are Dead* a character shouts at
the audience, "Fire!" He then quickly explains: "It's all right—I'm
demonstrating the misuse of free speech." Shouting "Fire!" in the
theater may well be the only jurisprudential analogy that has
assumed the status of a folk argument. A prominent historian
recently characterized it as "the most brilliantly persuasive expression that ever came from Holmes' pen." But in spite of its hallowed
position in both the jurisprudence of the First Amendment and the
arsenal of political discourse, it is and was an inapt analogy, even in
the context in which it was originally offered. It has lately become—despite, perhaps even because of, the frequency and promis-
cuousness of its invocation—little more than a caricature of logical
argumentation.

The case that gave rise to the "Fire!"-in-a-crowded-theater anal-   3
ogy—*Schenck* v. *United States*—involved the prosecution of Charles
Schenck, who was the general secretary of the Socialist Party in
Philadelphia, and Elizabeth Baer, who was its recording secretary.
In 1917 a jury found Schenck and Baer guilty of attempting to cause
insubordination among soldiers who had been drafted to fight in
the First World War. They and other party members had circulated
leaflets urging draftees not to "submit to intimidation" by fighting
in a war being conducted on behalf of "Wall Street's chosen few."

Schenck admitted, and the Court found, that the intent of the   4 pamphlets' "impassioned language" was to "influence" draftees to resist the draft. Interestingly, however, Justice Holmes noted that nothing in the pamphlet suggested that the draftees should use unlawful or violent means to oppose conscription: "In form at least [the pamphlet] confined itself to peaceful measures, such as a petition for the repeal of the act" and an exhortation to exercise "your right to assert your opposition to the draft." Many of its most impassioned words were quoted directly from the Constitution.

5   Justice Holmes acknowledged that "in many places and in ordinary times the defendants, in saying all that was said in the circular, would have been within their constitutional rights." "But," he added, "the character of every act depends upon the circumstances in which it is done." And to illustrate that truism he went on to say,

> The most stringent protection of free speech would not protect a man in falsely shouting fire in a theater, and causing a panic. It does not even protect a man from an injunction against uttering words that may have all the effect of force.

6   Justice Holmes then upheld the convictions in the context of a wartime draft, holding that the pamphlet created "a clear and present danger" of hindering the war effort while our soldiers were fighting for their lives and our liberty.

7   The example of shouting "Fire!" obviously bore little relationship to the facts of the Schenck case. The Schenck pamphlet contained a substantive political message. It urged its draftee readers to *think* about the message and then—if they so chose—to act on it in a lawful and nonviolent way. The man who shouts "Fire!" in a crowded theater is neither sending a political message nor inviting his listener to think about what he has said and decide what to do in a rational, calculated manner. On the contrary, the message is designed to force action *without* contemplation. The message "Fire!" is directed not to the mind and the conscience of the listener but, rather, to his adrenaline and his feet. It is a stimulus to immediate *action*, not thoughtful reflection. It is—as Justice Holmes recognized in his follow-up sentence—the functional equivalent of "uttering words that may have all the effect of force."

8   Indeed, in that respect the shout of "Fire!" is not even speech, in any meaningful sense of that term. It is a *clang* sound—the equivalent of setting off a nonverbal alarm. Had Justice Holmes been more honest about his example, he would have said that freedom of

speech does not protect a kid who pulls a fire alarm in the absence of a fire. But that obviously would have been irrelevant to the case at hand. The proposition that pulling an alarm is not protected speech certainly leads to the conclusion that shouting the word *fire* is also not protected. But the core analogy is the nonverbal alarm, and the derivative example is the verbal shout. By cleverly substituting the derivative shout for the core alarm, Holmes made it possible to analogize one set of words to another—as he could not have done if he had begun with the self-evident proposition that setting off an alarm bell is not free speech.

The analogy is thus not only inapt but also insulting. Most Americans do not respond to political rhetoric with the same kind of automatic acceptance expected of schoolchildren responding to a fire drill. Not a single recipient of the Schenck pamphlet is known to have changed his mind after reading it. Indeed, one draftee, who appeared as a prosecution witness, was asked whether reading a pamphlet asserting that the draft law was unjust would make him "immediately decide that you must erase that law." Not surprisingly, he replied, "I do my own thinking." A theatergoer would probably not respond similarly if asked how he would react to a shout of "Fire!" 9

Another important reason why the analogy is inapt is that Holmes emphasizes the factual falsity of the shout "Fire!" The Schenck pamphlet, however, was not factually false. It contained political opinions and ideas about the causes of the war and about appropriate and lawful responses to the draft. As the Supreme Court recently reaffirmed (in *Falwell* v. *Hustler*), "The First Amendment recognizes no such thing as a 'false' idea." Nor does it recognize false opinions about the causes of or cures for war. 10

A closer analogy to the facts of the Schenck case might have been provided by a person's standing outside a theater, offering the patrons a leaflet advising them that in his opinion the theater was structurally unsafe, and urging them not to enter but to complain to the building inspectors. That analogy, however, would not have served Holmes's argument for punishing Schenck. Holmes needed an analogy that would appear relevant to Schenck's political speech but that would invite the conclusion that censorship was appropriate. 11

Unsurprisingly, a war-weary nation—in the throes of a know- 12 nothing hysteria over immigrant anarchists and socialists—welcomed the comparison between what was regarded as a seditious political pamphlet and a malicious shout of "Fire!" Ironically, the "Fire!" analogy is nearly all that survives from the Schenck case; the ruling itself is almost certainly not good law. Pamphlets of the kind that resulted in Schenck's imprisonment have been circulated with impunity during subsequent wars.

13   Over the past several years I have assembled a collection of instances—cases, speeches, arguments—in which proponents of censorship have maintained that the expression at issue is "just like" or "equivalent to" falsely shouting "Fire!" in a crowded theater and ought to be banned, "just as" shouting "Fire!" ought to be banned. The analogy is generally invoked, often with self-satisfaction, as an absolute argument-stopper. It does, after all, claim the high authority of the great Justice Oliver Wendell Holmes. I have rarely heard it invoked in a convincing, or even particularly relevant, way. But that, too, can claim lineage from the great Holmes.

14   Not unlike Falwell, with his silly comparison between shouting "Fire!" and publishing an offensive parody, courts and commentators have frequently invoked "Fire!" as an analogy to expression that is not an automatic stimulus to panic. A state supreme court held that "Holmes' aphorism . . . applies with equal force to pornography"—in particular to the exhibition of the movie *Carmen Baby* in a drive-in theater in close proximity to highways and homes. Another court analogized "picketing . . . in support of a secondary boycott" to shouting "Fire!" because in both instances "speech and conduct are brigaded." In the famous Skokie case one of the judges argued that allowing Nazis to march through a city where a large number of Holocaust survivors live "just might fall into the same category as one's 'right' to cry fire in a crowded theater."

15   Outside court the analogies become even more badly stretched. A spokesperson for the New Jersey Sports and Exposition Authority complained that newspaper reports to the effect that a large number of football players had contracted cancer after playing in the Meadowlands—a stadium atop a landfill—were the "journalistic equivalent of shouting fire in a crowded theater." An insect researcher acknowledged that his prediction that a certain amusement park might become roach-infested "may be tantamount to shouting fire

in a crowded theater." The philosopher Sidney Hook, in a letter to the *New York Times* bemoaning a Supreme Court decision that required a plaintiff in a defamation action to prove that the offending statement was actually false, argued that the First Amendment does not give the press carte blanche to accuse innocent persons "any more than the First Amendment protects the right of someone falsely to shout fire in a crowded theater."

Some close analogies to shouting "Fire!" or setting off an alarm 16 are, of course, available: calling in a false bomb threat; dialing 911 and falsely describing an emergency; making a loud, gunlike sound in the presence of the President; setting off a voice-activated sprinkler system by falsely shouting "Fire!" In one case in which the "Fire!" analogy was directly to the point, a creative defendant tried to get around it. The case involved a man who calmly advised an airline clerk that he was "only here to hijack the plane." He was charged, in effect, with shouting "Fire!" in a crowded theater, and his rejected defense—as quoted by the court—was as follows: "If we built fireproof theaters and let people know about this, then the shouting of 'Fire!' would not cause panic."

Here are some more-distant but still related examples: the recent 17 incident of the police slaying in which some members of an onlooking crowd urged a mentally ill vagrant who had taken an officer's gun to shoot the officer; the screaming of racial epithets during a tense confrontation; shouting down a speaker and preventing him from continuing his speech.

Analogies are, by their nature, matters of degree. Some are closer 18 to the core example than others. But any attempt to analogize political ideas in a pamphlet, ugly parody in a magazine, offensive movies in a theater, controversial newspaper articles, or any of the other expressions and actions catalogued above to the very different act of shouting "Fire!" in a crowded theater is either self-deceptive or self-serving.

The government does, of course, have some arguably legitimate 19 bases for suppressing speech which bear no relationship to shouting "Fire!" It may ban the publication of nuclear-weapon codes, of information about troop movements, and of the identity of undercover agents. It may criminalize extortion threats and conspiratorial agreements. These expressions may lead directly to serious harm, but the mechanisms of causation are very different from that at

work when an alarm is sounded. One may also argue—less persua-
sively, in my view—against protecting certain forms of public
obscenity and defamatory statements. Here, too, the mechanisms of
causation are very different. None of these exceptions to the First
Amendment's exhortation that the government "shall make no law
. . . abridging the freedom of speech, or of the press" is anything like
falsely shouting "Fire!" in a crowded theater; they all must be
justified on other grounds.

20   A comedian once told his audience, during a stand-up routine,
about the time he was standing around a fire with a crowd of people
and got in trouble for yelling "Theater, theater!" That, I think, is
about as clever and productive a use as anyone has ever made of
Holmes's flawed analogy.

STUDY QUESTIONS

1. The key propositions in Holmes's argument are as follows:
   (1) Shouting "Fire!" in a theater creates a "clear and present
       danger."
   (2) Creating a "clear and present danger" is not protected by
       the First Amendment.
   (3) Shouting "Fire!" in a theater is not protected by the First
       Amendment.
   (4) Schenck's distributing the leaflet to draftees is similar to
       shouting "Fire!" in a theater.
   (5) Schenck's distributing the leaflet to draftees is not pro-
       tected by the First Amendment.
   Diagram the argument.
2. In an analogical argument, the first thing to check is the
   premise that makes the similarity claim—in this case, proposition (4). What similarities are there between Schenck's
   distributing the leaflet and shouting "Fire!" in a crowded
   theater?
3. In paragraphs 5 and 6, Dershowitz quotes Holmes as stating
   that the First Amendment does not protect words that cause
   a panic, "words that may have all the effect of force," or words
   that create a "clear and present danger." Given what you

know about Schenck's case, does it seem that his distributing the leaflet falls under any of these categories? If so, why? If not, what characteristic of Schenck's action prevents it from doing so?

4. Dershowitz thinks Holmes made a bad analogy, and he states what he thinks are two important disanalogies in paragraphs 7 and 10. What are they? How do they fit with your answers to question 3?

5. In paragraph 8, Dershowitz offers what he thinks is a better analogy for shouting "Fire!" in a theater; and in paragraph 11, he offers what he thinks is a better analogy for Schenck's handing out leaflets. Are they in fact better analogies?

Reference Chapters: 7, 16

PART NINE

On the Existence of God

E. K. DANIEL

Two Proofs of God's Existence

In the following selection, excerpted from a longer essay on proofs offered for the existence of God, Daniel presents and defends two traditionally powerful theistic arguments.

In section I, I said that my defense of theism will proceed by way   1
of defending the classical arguments for the existence of God. I also
said that my defense of the arguments would follow an earlier
defense of them by Julian Hartt.

[E. K. Daniel, excerpt from "A Defense of Theism," in E. D. Klemke, A. David Kline, and Robert Hollinger, eds., *Philosophy: The Basic Issues,* 3rd ed. New York: St. Martin's Press, 1990, pp. 269–272.]

2   Hartt's defense of these . . . arguments is extremely interesting and worthy of careful study. He mentions that: (a) there are several all-pervasive, broad characteristics of the world (universe)—or features of the world—which we experience; (b) each of the three *a posteriori* arguments focuses on one (or more) of these features; and (c) in each case, the feature(s) can only be explained by, and thus necessitate(s) the existence of, a being who is transcendent to the universe: God.

3   What are these features? And which of the three arguments focuses on which feature? We may put this in the form of a table:

Argument	*Feature (of universe)*
1. First-Cause (Cosmological)	finitude; contingency
2. Design (Teleological)	purposiveness; purposeful adaptation and arrangement

Again, in Hartt's view, these are objective features of the world (universe) which we experience, and each necessitates the existence of a (transcendent and infinite) God.

4   1. *The First-Cause (Cosmological) Argument.* The first-cause argument calls attention to and begins with the feature of *finitude and the related feature of contingency.* The argument also makes use of and rests on the notion of a *first cause* in connection with those features. Thus we find three variations. These different themes can give rise to several forms of the first-cause argument, or all of them can be included in a single argument. I propose to defend the cosmological argument by reformulating it as follows:

> Everything in the universe is finite.
>
> Whatever is finite is limited.
>
> Hence, whatever is limited cannot be the cause of its own existence.
>
> Everything in the universe is contingent.
>
> Whatever is contingent is dependent on something else for its existence.

Hence, whatever is contingent cannot be the cause of its own existence.

The totality of things making up the universe is also finite and contingent.

Thus, the totality (universe) must also have a cause for its existence.

Since it cannot be the cause of its own existence, the cause must be something external to the universe.

That is, since the universe cannot contain the reason for its existence within itself, the reason for its existence must be something external to it.

Hence, there must exist an infinite and self-subsistent (non-contingent) being who is the cause of the universe.

Unlike that which is finite and contingent, such a being must exist necessarily.

Such a being is commonly called God.

Therefore, there exists an infinite, necessary, and uncaused cause—God.

Someone may object: But why does the universe as a totality need a cause or explanation? Why can it not have existed infinitely in time? I reply: let us suppose it did. Then what the cosmological argument seeks is to provide answers to some questions: (1) Why does anything exist at all? (2) Why does it exist as it does rather than some other way? Or (1) Why is there a world at all? (2) Why is there this kind of universe rather than some other one? The answer is: Because of the purpose of an unlimited, infinite, and necessary being—an ultimate first cause which is itself uncaused—God.

In summary: Why does the fact that the universe is finite and contingent necessitate God's existence? The answer is that whatever is finite is limited. Hence, it cannot cause itself. And whatever is contingent is dependent on something else for its existence. Hence, it cannot be the cause of its own existence. The universe—as well as everything in it—is finite and contingent. There is nothing about it to indicate that it could be the cause of its own existence. Since it cannot be the cause of its own existence, it must have been caused by something else—a being external to the universe. Therefore, such a being must exist: an unlimited, necessary, and uncaused cause—God.

7   2. *The Design (Teleological) Argument.* The design argument
calls attention to another feature of the universe, that of *purposive-*
ness, or purposeful adaptation of means to ends.

8   The word *teleological* comes from the Greek word *telos,* meaning
purpose or goal. Theism is a teleological metaphysics through and
through. Hence, it is understandable and natural for there to be
arguments which focus on the notion of purpose. Among them is
the teleological argument for the existence of God. This argument
claims that the many features of design, purpose, and adaptation in
the universe are indications of a Cosmic Intelligence or Mind—
God—which designed, planned, and brought the universe into
existence, I propose to defend the teleological argument by reformulating it as follows:

> Suppose that while walking along an ocean beach, or a barren field,
> we come upon an object, such as a watch.
>
> If we examine the watch, we find that it shows evidence of purpose
> and design.
>
> We detect orderliness and intricacy.
>
> We find an adaptation of means to ends (the parts are arranged to
> work together to enable the hands to move and to enable us to tell
> time).
>
> All of this is evidence of rationality and design.
>
> Hence, there exists a rational being who designed and brought the
> watch into being.

Similarly:

> Look out at the universe and the things within it.
>
> The universe also shows evidence of design and purpose.
>
> We detect orderliness and intricacy.
>
> Moreover, we find a marvelous adaptation of means to ends.
>
> An example of such adaptation is the existence of two sexes for the
> end of procreation or the structure of the eye for the end of seeing.
>
> All this is also evidence of rationality and design.
>
> Hence, there must exist a rational being who designed and brought
> the universe into existence.
>
> That is, there must exist a Cosmic Designer—God.

It may be objected: But could not the universe have resulted from   9
chance? I reply: Although there may be chance *in* the universe, the
universe *itself* is not the product of chance but of *intelligent purpose.*
The environment in which we find ourselves is not a fortuitously
functioning mechanism; nor is it an organism. It is imbued with
purpose. Everyone admits that humans show evidence of mind and
purposive behavior—as in designing and making a house. But we
cannot suppose that purposeful activity is limited to humans and
that everything else in nature is blind or the result of sheer chance.
Why not? Because *our* minds, our intelligent planning, have not
made the universe. Therefore, there must exist a being who designed the universe and brought it into existence.

In summary: Why does the fact that the universe is purposeful   10
necessitate God's existence? The reason is that whatever is purposeful shows signs of intelligence—mind. Hence, what is purposive
cannot have come about accidentally, or from something non-pur-
posive. Hence, the only way to explain the purposiveness in the
universe is: It got here because of the thought, design, and activity
of a Cosmic Intelligence—God.

STUDY QUESTIONS

1. The first step of the First Cause argument goes like this:

   Everything in the universe is finite.

   Whatever is finite is limited.

   Hence, whatever is limited cannot be the cause of its own
   existence.

   What assumed premise does this argument depend on?
2. For each of the remaining steps in the First Cause argument,
   determine whether they depend on assumed premises.
3. The design argument proceeds by making an analogy from
   a watch and the universe. How many points of similarity
   between the watch and the universe does the analogy depend
   on?

4. The design argument can be diagrammed as follows:

(1) The watch is   +   (2) The watch requires an
     orderly   intelligent desiger

_____

(3)   +   (4) The universe is orderly

_____

(5) The universe requires an intelligent designer.

What inductive generalization is required at (3) in order to complete the argument?

5. Does it make sense to suppose that if God exists, he must be orderly and intricate, and his parts must show an adaption of means to ends? If so, does the design argument lead us to suppose that God must have been designed by a rational being?

Reference Chapters: 7, 10, 11

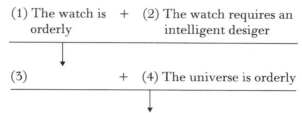

GEORGE H. SMITH

God Does Not Exist

George H. Smith is a senior research fellow at the Institute for Humane Studies at George Mason University in Virginia. He

[George H. Smith, excerpt from Chapter 3 of *Atheism: The Case Against God.* Buffalo, N.Y.: Prometheus Books, 1979, pp. 81–87, references deleted.]

is the author of Atheism, Ayn Rand, and Other Heresies, *and*
Atheism: The Case Against God, *from which the following is
excerpted.*

Briefly, the problem of evil is this: If God does not know there is   1
evil, he is not omniscient. If God knows there is evil but cannot
prevent it, he is not omnipotent. If God knows there is evil and can
prevent it but desires not to, he is not omnibenevolent. If, as the
Christian claims, God is all-knowing and all-powerful, we must
conclude that God is not all-good. The existence of evil in the
universe excludes this possibility.

There have been various attempts to escape from the problem of   2
evil, and we shall briefly consider the more popular of these. But
one point requires emphasis. The Christian, by proclaiming that
God is good, commits himself to the position that man is capable of
distinguishing good from evil—for, if he is not, how did the Christian arrive at his judgment of "good" as applied to God? Therefore,
any attempt to resolve the problem of evil by arguing that man
cannot correctly distinguish good from evil, destroys the original
premise that it purports to defend and thus collapses from the
weight of an internal inconsistency. If the human standards of good
and evil are somehow invalid, the Christian's claim that God is good
is equally invalid.

One general theological approach to the problem of evil consists   3
of the claim that evil is in some way unreal or purely negative in
character. This argument, however, is so implausible that few
Christians care to defend it. The first problem with it, as Antony
Flew notes, is: "If evil is really nothing then what is all the fuss
about sin about: nothing?"

In *Some Dogmas of Religion*, John McTaggart quickly disposes   4
of the claim that evil is in some way unreal:

> Supposing that it could be proved that all that we think evil was in
> reality good, the fact would still remain that we think it evil. This
> may be called a delusion or a mistake. But a delusion or mistake is
> as *real* as anything else. A savage's erroneous belief that the earth is
> stationary is just as real a fact as an astronomer's correct belief that
> it moves. The delusion that evil exists, then, is real. But then . . . it
> seems certain that a delusion or an error which hid from us the
> goodness of the universe would itself be evil. And so there would be

> real evil after all. . . . However many times we pronounce evil unreal, we always leave a reality behind, which in its turn is to be pronounced evil.

5   As for the argument that evil is purely negative, a privation of the good (as disease may be said to be the absence of health), Wallace Matson provides this illuminating example in *The Existence of God:*

> It may console the paralytic to be told that paralysis is mere lack of mobility, nothing positive, and that insofar as he *is,* he is perfect. It is not clear, however, that this kind of comfort is available to the sufferer from malaria. He will reply that his trouble is not that he lacks anything, but rather that he has too much of something, namely, protozoans of the genus *Plasmodium.*

6   Any attempt to absolve God of the responsibility for evil by claiming that, in the final analysis, there is no such thing as evil is, as Matson puts it, "an unfunny joke." This approach merely ends up by negating our human standards of good and evil, which, as previously indicated, undercuts the argument at its root.

7   Another common effort to reconcile God and evil is to argue that evil is the consequence of man's freely chosen actions. God, through his gift of free will, gave man the ability to distinguish and choose between good and evil, right and wrong. As a free agent, man has the potential to reach a higher degree of perfection and goodness than if he were a mere robot programmed to behave in a given manner. Thus it is good that man has free will. But this entails the opportunity for man to select evil instead of good, which has been the case in the instances of torture, murder, and cruelty which some men inflict upon others. The responsibility for these actions, however, rests with man, not with God. Therefore, concludes the Christian, evil does not conflict with the infinite goodness of God.

8   While this approach has some initial plausibility, it falls far short of solving the problem of evil. We are asked to believe that God created man with the power of choice in the hope that man would voluntarily pursue the good, but that man thwarts this desire of God through sin and thus brings evil upon himself. But, to begin with, to speak of frustrating or acting contrary to the wishes of an omnipotent being makes no sense whatsoever. There can be no barriers to divine omnipotence, no obstacles to thwart his desires, so

we must assume that the present state of the world is precisely as God desires it to be. If God wished things to be other than they are, nothing could possibly prevent them from being other than they are, man's free will notwithstanding. In addition, we have seen that free will is incompatible with the foreknowledge possessed by an omniscient being, so the appeal to free will fails in this respect as well. In any case, God created man with full knowledge of the widespread suffering that would ensue, and, given his ability to prevent this situation, we must presume that God desired and willed these immoral atrocities to occur.

It is unfair to place the responsibility for immoral actions on man's free will in general. Individual men commit atrocities, not the bloodless abstraction "man." Some men commit blatant injustices, but others do not. Some men murder, rob, and cheat, but others do not. Some men choose a policy of wanton destructiveness, but others do not. And we must remember that crimes are committed by men against other men, innocent victims, who cannot be held responsible. The minimum requirement for a civilized society is a legal system whereby the individual liberties of men are protected from the aggressive activities of other men. We regard the recognition and protection of individual rights as a moral necessity, and we condemn governments that fail to provide a fair system of justice. How, then, are we to evaluate a God who permits widespread instances of injustice when it is easily within his power to prevent them? The Christian believes in a God who displays little, if any, interest in the protection of the innocent, and we must wonder how such a being can be called "good." 9

The standard reply to this objection is that God rewards the virtuous and punishes the wicked in an afterlife, so there is an overall balance of justice. An extreme variation of this tactic was reported in *The New York Times* of September 11, 1950. Referring to the Korean War, this article states: "Sorrowing parents whose sons have been drafted or recalled for combat duty were told yesterday in St. Patrick's Cathedral [by Monsignor William T. Greene] that death in battle was part of God's plan for populating 'the kingdom of heaven.'" 10

This approach is so obviously an exercise in theological rationalization that it deserves little comment. If every instance of evil is to be rectified by an appeal to an afterlife, the claim that God is 11 all-good has no relevance whatsoever to our present life. Virtually any immoral action, no matter how hideous or atrocious, can be explained away in this fashion—which severs any attempt to discuss the alleged goodness of a creator from reference to empirical evidence. More importantly, no appeal to an afterlife can actually eradicate the problem of evil. An injustice always remains an injustice, regardless of any subsequent efforts to comfort the victim. If a father, after beating his child unmercifully, later gives him a lollipop as compensation, this does not erase the original act or its evil nature. Nor would we praise the father as just and loving. The same applies to God, but even more so. The Christian may believe that God will punish the perpetrators of evil and compensate the victims of injustice, but this does not explain why a supposedly benevolent and omnipotent being created a world with evildoers and innocent victims in the first place. Again, we must assume that there are innocent victims because God desires innocent victims; from the standpoint of Christian theism, there is simply no other explanation. If an omnipotent God did not want innocent victims, they could not exist—and, by human standards, the Christian God appears an immoral fiend of cosmic dimensions.

12   Even if we overlook the preceding difficulties, the appeal to free will is still unsuccessful, because it encompasses only so-called *moral* evils (*i.e.,* the actions of men). There remains the considerable problem of *physical* evils, such as natural disasters, over which man has no control. Why are there floods, earthquakes and diseases that kill and maim millions of persons? The responsibility for these occurrences obviously cannot be placed on the shoulders of man. From an atheistic standpoint, such phenomena are inimical to man's life and may be termed evil, but since they are the result of inanimate, natural forces and do not involve conscious intent, they do not fall within the province of moral judgment. But from a Christian perspective, God—the omnipotent creator of the natural universe—must bear ultimate responsibility for these occurrences, and God's deliberate choice of these evil phenomena qualifies him as immoral.

13   There is an interesting assortment of arguments designed to explain the existence of natural evils. Some theologians argue that evil exists for the sake of a greater good; others maintain that apparent evils disappear into a universal harmony of good. Al-

though something may appear evil to man, we are assured by the Christian that God is able to view the overall perspective, and any apparent evil always turns out for the best. These approaches share the premise that man cannot understand the ways of God, but this simply pushes us into agnosticism. It will not do for the Christian to posit an attribute of God and, when asked to defend that attribute, contend that man cannot understand it.

If we are incorrect in calling natural disasters, diseases and other 14 phenomena evil, then man is incapable of distinguishing good from evil. But if this is the case, by what standard does the Christian claim that God is good? What criterion is the Christian using?

If man cannot pass correct moral judgments, he cannot validly 15 praise *or* condemn anything—including the Christian God. To exclude God from the judgment of evil is to exclude him from the judgment of good as well; but if man can distinguish good from evil, a supernatural being who willfully causes or permits the continuation of evil on his creatures merits unequivocal moral condemnation.

Some Christians resort to incredible measures to absolve their 16 God from the responsibility for evil. Consider this passage from *Evil and the God of Love* in which John Hick attempts to reconcile the existence of an omnibenevolent deity with the senseless disasters that befall man:

> . . . men and women often act in true compassion and massive generosity and self-giving in the face of unmerited suffering, especially when it comes in such dramatic forms as an earthquake or a mining disaster. It seems, then, that in a world that is to be the scene of compassionate love and self-giving for others, suffering must fall upon mankind with something of the haphazardness and inequity that we now experience. It must be apparently unmerited, pointless, and incapable of being morally rationalized. For it is precisely this feature of our common human lot that creates sympathy between man and man and evokes the unselfish kindness and goodwill which are among the highest values of personal life.

Aside from displaying a low regard for man's "highest values" 17 and their origins, Hick illustrates an important point: *There is virtually nothing which the Christian will accept as evidence of God's evil.* If disasters that are admittedly "unmerited, pointless, and incapable of being morally rationalized" are compatible with the "goodness" of God, what could possibly qualify as contrary evidence? The "goodness" of God, it seems, is compatible with any conceivable state of affairs. While we evaluate a man with reference to his actions, we are not similarly permitted to judge God. God is immune from the judgment of evil as a matter of principle.

18   Here we have a concrete illustration of theological "reasoning." Unlike the philosopher, the theologian adopts a position, a dogma, and then commits himself to a defense of that position come what may. While he may display a willingness to defend this dogma, closer examination reveals this to be a farce. His defense consists of distorting and rationalizing all contrary evidence to meet his desired specifications. In the case of divine benevolence, the theologian will grasp onto any explanation, no matter how implausible, before he will abandon his dogma. And when finally pushed into a corner, he will argue that man cannot understand the true meaning of this dogma.

19   This brings us to our familiar resting place. The "goodness" of God is different in kind from goodness as we comprehend it. To say that God's "goodness" is compatible with the worst disasters imaginable, is to empty this concept of its meaning. By human standards, the Christian God cannot be good. By divine standards, God may be "good" in some unspecified, unknowable way—but this term no longer makes any sense. And so, for the last time, we fail to comprehend the Christian God.

STUDY QUESTIONS

1. In the first paragraph, Smith presents a brief account of the problem of evil. As stated, the argument relies on some assumed premises. Rewrite the argument, filling in the assumed premises, and determine whether it is valid.

2. For the rest of the selection Smith proceeds by considering a number of attempts to reconcile the existence of the Christian God with the existence of evil, and arguing that they all fail. How many distinct attempts at reconciliation does Smith consider? Can you think of any others?

3. In paragraph 18, Smith concludes that theologians who defend the Christian God against the problem of evil are not really reasoning but rather "rationalizing all contrary evidence." Judging from the strength of the various responses to the problem of evil, how much evidence does Smith have for this claim?

Reference Chapters: 7, 10

PART TEN

Poverty, Crime, and Gun Control

DON B. KATES, JR.

Handgun Bans: Facts to Fight With

Donald Kates, Jr., is an attorney, the author of Firearms and
Violence: Issues of Public Policy, *and a leading advocate of
the right to own firearms. In the following selection Kates defends handgun ownership on a number of grounds, one of
which is the surprising claim that gun ownership may reduce
crime.*

The handgun debate has produced a plethora of emotional rhetoric   1
on both sides of the issue, but very little hard research. On one side
are the emotional bumper-sticker slogans full of patriotic posturing.

[Don B. Kates, Jr., "Handgun Bans—Facts to Fight With," in *Guns and Ammo
Annual, 1984.* Los Angeles: Petersen, 1984, pp. 4, 6, 8–11.]

On the other are equally emotional and sensationalized horror stories of innocent citizens killed by handguns, supplemented by supposedly neutral but, in fact, "result-oriented" social science research that is either misleading or downright inaccurate.

2   Gun owners nod enthusiastically at "Guns Don't Kill People, People Do" and "When Guns are Outlawed, Only Outlaws Will Have Guns," but such cliches are virtually useless in intelligent debate with someone who is not committedly pro-gun. Guns *do* kill people, just as knives and hand grenades do. If they didn't kill or injure or at least present that threat, they would be useless as instruments of self-defense. And even if "Only Outlaws Will Have Guns" if guns are outlawed, anti-gun forces have made it clear that they are willing to tolerate firearms possession by hardened criminals if a handgun ban would result in disarming the self-protection owner whom they believe responsible for murder.

3   If this characterization of murderers were accurate, banning handguns would seem an appealingly simple means of reducing domestic and acquaintance homicide. Most killings are not, however, perpetrated by the average noncriminal citizen whose law-obedient mentality (it is believed) would induce him to give up handguns in response to a ban. Refuting "the myth that the typical homicide offender is just an ordinary person" with "no previous criminal record," Professor Gary Kleck of Florida State University's School of Criminology notes F.B.I. figures showing two-thirds of all murderers to have previous felony conviction records.

4   Moreover, a murderer's prior arrest record is likely to substantially underrepresent the real prior violence history. Unlike robbers, who generally strike at strangers, murderers' prior violence may have been directed against relatives or acquaintances, that is, the same kinds of people murderers end up killing. Such prior violent incidents may have never led to arrest or conviction, either because the victim did not press charges or because the police refuse to interfere in "a family affair." A study in Kansas City revealed that in 85 percent of domestic homicide cases, the police have had to be summoned to the home at least once before the killing occurred, and in 50 percent of the cases, the police had stopped beatings five or more times before the actual murder. In short, these people are criminals no less hardened than the professional robbers whom everyone agrees a handgun ban won't disarm. Unlike the average

citizen, the typical murderer will not scruple to keep his gun in spite of a ban.

Unfortunately, the quality of most research associated with handgun control has been on par with the sort found in UFO magazines. Another example is the tired old line about gun controls working in Europe and Japan. The gun bans of the European countries commonly compared to the U.S. were not enacted to reduce general violence (with which those countries have been little affected), but were enacted to prevent the assassinations and political terrorism from which England, Germany, and so forth, still suffer far more than we.

In fact, prohibitionists abruptly stopped referring to England in 1971 with the appearance from Cambridge University of the first in-depth study of that country's handgun permit law. This Cambridge study attributes England's comparatively low violence wholly to cultural factors, pointing out that until 1920 England had far fewer gun controls than most American states. Yet England had far less violence at that time than did those states or than England now has. Those who blame greater handgun availability for our greater rates of handgun homicide ignore the fact that rates of murder with knives or without any weapon (i.e., with hands and feet) are also far lower in England. The study's author rhetorically asks whether it is claimed that knives are less available in England than in the U.S. or that the English have fewer hands and feet than Americans. As a subsequent British government publication puts it, although "one reason often given for American homicide is the easy availability of firearms . . . the strong correlation with racial and linked socio-economic variables suggests that the underlying determinants of the homicide rate relate to particular cultural factors."

European comparisons would be incomplete without mention of Switzerland, where violence rates are very low though every man of military age is required to own a handgun or fully automatic rifle. Israeli violence is similarly low, though the populace is even more intensively armed. A comparison with handgun-banning Japan's low homicide rate is plainly inappropriate because of our vastly different culture and heritage and our substantial ethnic heterogeneity. (The only valid comparison reinforces the irrelevancy of gun bans: It is that Japanese-Americans, with full access to handguns, have a slightly lower homicide rate than their gunless counterparts in Japan.) An appropriate comparison to Japan would be Taiwan. Despite even more stringent anti-handgun laws, it has a homicide rate greater than ours and four times greater than Japan's. Similarly the U.S. might well be compared to South Africa, a highly industrialized and ethnically heterogeneous country. Despite one of the world's most stringent "gun control" programs, South Africa's homicide rate, factoring out politically associated killings, is twice ours.

8   The moral of the story is that nothing about the correlation between levels of handgun ownership and violent crimes could lead one to conclude there is a cause and effect relationship. But it is simply taken for granted by many that there *is* a relationship, and they cite only those countries that have lower crime rates than America to "prove" it.

9   Writers on both sides of the barricade have too often started with conclusions and worked to justify them. For that reason, much of the best research has come not from conservative sources that have traditionally supported the right to own handguns, but from those who have converted to a position favorable to handgun ownership and feel a need to explain their aberrant positions.

10   In order to understand the preponderance of misinformation in the handgun ownership debate, it is necessary to trace some of the ideas that the anti-gun movement has used to justify its position. In my book *Firearms and Violence: Issues of Public Policy* (San Francisco: Pacific Institute for Public Policy Research, 1983), Gary Kleck of the Florida State University School of Criminology and David Bordua of the University of Illinois, Urbana, Department of Sociology have identified what they call "the key assumptions of gun control." The first we will consider for this article is:

People who buy guns for self-defense are the victims of self-deception and a mistaken belief in the protective efficacy of gun ownership. In fact, guns are useless for self-defense or protection of a home or business.

11   Fundamental to systematic discussion of these issues is the distinction between the self-defense value that gun ownership may have and its crime deterrence value. Anti-gun lobbyists are unassailably correct in asserting that a gun owner rarely has the opportunity to defend a home or business against burglars who generally take pains to strike only at unoccupied premises. But this fails to

address two important issues of deterrence. Kleck and Bordua calculate that a burglar's small chance of being confronted by a gun-armed defender probably exceeds that of his being apprehended, tried, convicted and actually serving any time. Which, they ask, is a greater deterrent: a slim chance of being punished or a slim chance of being shot?

Even more important, fear of meeting a gun-armed defender may be one factor in the care most burglars take to strike at only unoccupied premises. In this connection, remember that it is precisely because burglary is generally a nonconfrontation crime that victim injury or death is so very rarely associated with it—in contrast to robbery, where victim death is an all too frequent occurrence. If the deterrent effect of victim gun possession helps to make burglary an overwhelmingly nonconfrontational crime, thereby minimizing victim death or injury, that effect *benefits* burglary victims generally, even if the gun owners gain no particular self-defense value thereby.

The most recent evidence of the deterrent value of gun ownership appears in a survey taken in 10 major prisons across the United States by the Social and Demographic Research Institute of the University of Massachusetts. It confirms earlier prison surveys in which inmates stated that (1) they and other criminals tried to avoid victims they believed may have been armed and that (2) they favor gun prohibition because, by disarming the victim, it will make their lives safer without affecting their access to illegal guns.

Increasingly, police are concluding and even publicly proclaiming that they cannot protect the law-abiding citizens and that it is not only rational to choose to protect oneself with firearms but societally beneficial because it deters violent crimes. Because of the lack of coherent evidence on the subject until the late 1970s, such views necessarily were only intuitively or anecdotally based. They were controverted by the citation of isolated, artificially truncated statistics supposedly showing that citizens rarely are able to kill criminals in self-defense. In fact, civilians justifiably kill about as many violent criminals across the nation as do the police. In California and Miami, private citizens kill twice as many criminals; in Chicago and Cleveland, three times as many.

Even when accurately reported, such statistics are unfair in that they underrepresent the full self-defense value of guns; as with the

12

13

14

15

police, the measure of the success of armed citizens lies not in the number of criminals they kill, but in the total number whom they defeat by wounding, driving off, or arresting. In his paper for *Firearms and Violence,* Professor Wright concludes that incidents of people defending themselves with handguns are even more numerous than incidents of handgun misuse by criminals against citizens. In other words, while there are all too many crimes committed with handguns, there are even more crimes being foiled by law-abiding gun owners.

16 In his extensive study of Atlanta data, Philip Cook concluded that a robber's chance of dying in any one year in that city is doubled by committing only seven robberies because of the risk of counterattack by a potential victim. In addition to illustrating the self-defense value of handgun ownership, that is a pretty good indication of the kind of deterrent effect caused by gun ownership. Cook's work also shows that areas with the strongest anti-gun laws have the highest rates of crime. That in itself does not prove anything except the spurious nature of the data distributed by anti-gun forces that purports to prove the opposite. Since high crime rates often lead governments to seek solutions in gun bans, the correlation between gun control and high crime rates does not necessarily prove that bans cause violence. But more meaningful correlations can be found when examining the opposite case, where the public has actually *increased* the level of gun ownership.

17 The Orlando Police Department, when plagued with a sharply rising number of rapes in the city, undertook a firearms training program for women between October 1966 and March 1967. Kleck and Bordua studied the effects of the program and found that the rape rate in Orlando fell by almost 90 percent from a level of 35.91 per 100,000 inhabitants in 1966 to 4.18 in 1967. The surrounding areas and Florida in general experienced either constant or increasing rape rates during the same period, as did the United States in general. Another benefit from the program seems to have been the corresponding decrease in Orlando burglaries. Though the rape-protection program was well-publicized, the anti-burglary aspect of gun ownership and training was not emphasized in the press. Burglars apparently made a connection between women's willingness and capability to defend their bodies and the increased risk of taking their property.

18 Similar programs have resulted in decreasing store robberies by

as much as 90 percent in Highland Park and Detroit, Michigan, and New Orleans. Perhaps the most publicized is that of Kennesaw, Georgia, where the city council passed a kind of reverse gun control act requiring citizens to keep guns in their homes. In the ten months that followed, there was an 89 percent decrease in burglary.

These programs have been criticized on the theory that the crime rate has not been lowered but simply shifted to other areas. This does not appear to have been the case in Orlando. There rape fell by 9 percent in the surrounding communities even as it fell by 90 percent in Orlando in the first year after the handgun training program was publicized. Also, criticisms that gun ownership is causing crime to shift to nongun-owning areas are based on the perverse idea that defending oneself from rape or burglary is really an offense against others who do not.

It is not likely that women who are willing and able to protect themselves from rapists would have much sympathy with that view, even if it were true. To take the argument to its logical conclusion, one would have to admit that anybody who does anything to discourage crime (locking doors, calling for help, summoning the police, or staying indoors and out of alleys at night) is endangering the others who will not.

The evidence does show gun ownership acts as a deterrent to criminal activity for those who own them *and* those who do not. Opponents of handgun ownership get around that by saying:

Handguns are more dangerous to their owners and the family and friends of the owners than they are to criminals. Handguns kill more people through accident and criminal assault than they save.

Once again, there have been several widely quoted studies supporting this view that have become part of the conscious and subconscious rationale for banning handguns. One study, by Rushforth, purported to show that six times as many Cleveland householders died in gun accidents as killed burglars. It was discredited by Professor James Wright of the University of Massachusetts at Amherst in his paper for the 1981 annual meeting of the American Society of Criminology. Research done with San Francisco City Supervisor Carol Ruth Silver has indicated that between 1960 and 1975 the number of instances where handguns were used for defense exceeded the cases where they were misused to kill by a ratio of 15 to 1.

According to a monitoring of 42 of the nation's largest newspapers between June 1975 and July 1976, 68 percent of the time that police used firearms, they successfully prevented a crime or caught the criminal. On the other hand, the success rate for private citizens was 83 percent. This is not particularly surprising since the police must usually be summoned to the scene of criminal activity while the private gun owner is more likely to be there when it occurs. At the same time, Kleck and Bordua conclude that citizens who resist crimes with weapons are much less likely to be injured than those who attempt to resist without weapons.

24   Anti-gun forces cover themselves in the question of self-defense by asserting that the absence of handguns will lead criminals to use other less dangerous weapons, lowering the death and injury rate in confrontations with criminals. The heart of the argument, as stated by Kleck and Bordua, is:

Guns are five times deadlier than the weapons most likely to be substituted for them in assaults where firearms are not available.

25   To begin with, the above statement does not differentiate between different types of guns. Most anti-gun measures are aimed at handguns because banning all guns is politically unlikely. Furthermore, the statement is based on the assumption that the second choice for a criminal who is denied a handgun would be a knife. In fact, a Massachusetts University survey of 10 major prisons indicates that in the majority of cases, a criminal denied access to a handgun will turn to a sawed-off shotgun or long rifle. Approximately 50 percent of all criminals and 75 percent of those with a history of handgun use said they would use a sawed-off shotgun or rifle. While handguns are slightly more deadly than knives, shotguns and rifles are three to eight times as deadly as handguns. If handguns disappeared but only 19 percent of criminals turned to long guns, the same amount of fatalities could be expected to occur. If 50 percent used shotguns or rifles, there would actually be an increase in assault-related deaths by as much as 300 percent.

26   Another danger associated with the upgrading of firearms due to a handgun ban is the increased risk of accidental wounds and fatalities due to the longer range of shotguns and rifles. This would be the likely consequence of a measure banning handguns but allowing the ownership of long guns for self-defense. A shot from a handgun that misses its target will usually come to rest in a wall much sooner than a shotgun or rifle blast, which is much more likely

to continue on to impact whatever is on the other side of the wall. The danger of accidental death increases enormously when long guns replace handguns as the arm kept for home defense. Even now, long guns are involved in 90 percent of all accidental firearms deaths, though they probably represent less than 10 percent of the guns kept loaded at any one time.

So we see, handgun ownership is more than a civil liberty, it is valuable for society as a whole. Given the obvious nature of the analyses, it is surprising that the anti-gun forces in this country have succeeded to the degree they have. 27

Recent court decisions have made it clear that the government has no responsibility to protect the citizenry. In an important case, *Warren* vs. *District of Columbia,* three women brought suit against the local government because of lack of police action. Two men broke into their home and found one of the women. The other two were upstairs and called the police twice over a period of half an hour. After the woman downstairs had been beaten, raped and sodomized into silence, her roommates believed that the police had arrived. They went downstairs and "for the next 14 hours the women were held captive, raped, robbed, beaten, forced to commit sexual acts upon each other, and made to submit to the sexual demands of [their attackers]." The three women lost their suit and an appeal because, as the courts universally hold, "a government and its agents are under no general duty to provide public services, such as police protection to any particular citizen." 28

This incident and many others like it took place in the city with the most stringent anti-gun law in the country. The D.C. law required handgun owners to register their guns and then disallowed the ownership of any new arms. Furthermore, the law made it illegal to keep a firearm assembled or loaded in the home for self-defense. There is a real ideological question as to whether a government can disclaim responsibility to protect its citizens *and* take away their means to protect themselves. Pragmatically though, one must admit that the government is basically unable to protect every citizen at all times, regardless of the legal position. 29

What good are handguns? The evidence has led many to believe that they are the largest single deterrent against crime. A recent study conducted by the Boston Police Department showed that the majority of high-ranking police administrators and police chiefs 30 across the nation actually favor allowing law-abiding citizens to *carry* guns for self-protection. What can be done about the well-intentioned but misinformed foes of the handgun? Perhaps the most important thing would be the self-education of those who already defend that right. Only with intelligent, informed argument can the gun banners be convinced of the foolishness of their position. It is too easy to blame the media for bias or lack of information. Those who hold a position also have a responsibility to be informed and put those arguments forward. I hope that this article will help to do that.

STUDY QUESTIONS

1. Kates's overall conclusion is that handguns should not be banned. In defending this conclusion, he argues five major claims:
   (1) Handgun bans will not reduce crime;
   (2) Cultural factors, not the legality of handguns, are responsible for the United States' crime rate;
   (3) Handguns have a self-defense value and a deterrence value;
   (4) Handguns are not more dangerous to their owners, friends, and families than they are to criminals;
   (5) If handguns are banned, gun fatalities are likely to increase.
   Identify the major premises Kates uses to support each claim.
2. In paragraphs 5 through 8, Kates is arguing for claim (2), i.e., that comparing the United States with other countries does not help the anti-gun position. Find an example of a negative use of Mill's Methods in this argument.
3. In paragraph 17, Kates is arguing in support of claim (3) by arguing that a firearms-training program for women in Orlando was causally responsible for the decline in the rape rate the following year. Which of Mill's Methods is Kates using in this argument?
4. At the end of paragraph 25, Kates states: "If 50 percent [of criminals] used shotguns or rifles, there would actually be an

increase in assault-related deaths by as much as 300 percent."
Is this a statistic Kates cites, or is it the conclusion of an
argument?

5. In the essay in favor of banning handguns that follows, do
John Henry Sloan, et al., raise any arguments that Kates has
not addressed in this essay?

Reference Chapters: 15, 17

JOHN HENRY SLOAN, ET AL.

Handgun Regulations, Crime, Assaults, and Homicide: A Tale of Two Cities

*In the following article, a team of medical doctors uses the re-
sults of a comparative study of Seattle, Washington, and Vancouver, British Columbia, to argue that good empirical evidence exists to support the conclusion that banning handguns
will reduce crime.*

Abstract: To investigate the associations among handgun regulations, assault and other crimes, and homicide, we studied robberies,
burglaries, assaults, and homicides in Seattle, Washington, and
Vancouver, British Columbia, from 1980 through 1986.

Although similar to Seattle in many ways, Vancouver has

1

2

[John Henry Sloan, Arthur L. Kellermann, Donald T. Reay, James A. Ferris,
Thomas Koepsell, Frederick P. Rivara, Charles Rice, Laurel Gray, and James
LoGerfo, "Handgun Regulations, Crime, Assaults, and Homicide: A Tale of Two
Cities." *New England Journal of Medicine* 319, November 10, 1988, pp. 1256–
1262.]

adopted a more restrictive approach to the regulation of handguns. During the study period, both cities had similar rates of burglary and robbery. In Seattle, the annual rate of assault was modestly higher than that in Vancouver (simple assault: relative risk, 1.18; 95 percent confidence interval, 1.15 to 1.20; aggravated assault: relative risk, 1.16; 95 percent confidence interval, 1.12 to 1.19). However, the rate of assaults involving firearms was seven times higher in Seattle than in Vancouver. Despite similar overall rates of criminal activity and assault, the relative risk of death from homicide, adjusted for age and sex, was significantly higher in Seattle than in Vancouver (relative risk, 1.63; 95 percent confidence interval, 1.28 to 2.08). Virtually all of this excess risk was explained by a 4.8-fold higher risk of being murdered with a handgun in Seattle as compared with Vancouver. Rates of homicide by means other than guns were not substantially different in the two study communities.

3   We conclude that restricting access to handguns may reduce the rate of homicide in a community.

4   Approximately 20,000 persons are murdered in the United States each year, making homicide the 11th leading cause of death and the 6th leading cause of the loss of potential years of life before age 65.[1–3] In the United States between 1960 and 1980, the death rate from homicide by means other than firearms increased by 85 percent. In contrast, the death rate from homicide by firearms during this same period increased by 160 percent.[3]

5   Approximately 60 percent of homicides each year involve firearms. Handguns alone account for three fourths of all gun-related homicides.[4] Most homicides occur as a result of assaults during arguments or altercations; a minority occur during the commission of a robbery or other felony.[2,4] Baker has noted that in cases of assault, people tend to reach for weapons that are readily available.[5] Since attacks with guns more often end in death than attacks with knives, and since handguns are disproportionately involved in intentional shootings, some have argued that restricting access to handguns could substantially reduce our annual rate of homicide.[5–7]

6   To support this view, advocates of handgun control frequently cite data from countries like Great Britain and Japan, where the rates of both handgun ownership and homicide are substantially lower than those in the United States.[8] Rates of injury due to assault in Denmark are comparable to those in northeastern Ohio, but the

Danish rate of homicide is only one fifth as high as Ohio's.[5,6] In Denmark, the private ownership of guns is permitted only for hunting, and access to handguns is tightly restricted.[6]

Opponents of gun control counter with statistics from Israel and Switzerland, where the rates of gun ownership are high but homicides are relatively uncommon.[9] However, the value of comparing data from different countries to support or refute the effectiveness of gun control is severely compromised by the large number of potentially confounding social, behavioral, and economic factors that characterize large national groups. To date, no study has been able to separate the effects of handgun control from differences among populations in terms of socioeconomic status, aggressive behavior, violent crime, and other factors.[7] To clarify the relation between firearm regulations and community rates of homicide, we studied two large cities in the Pacific Northwest: Seattle, Washington, and Vancouver, British Columbia. Although similar in many ways, these two cities have taken decidedly different approaches to handgun control.

7

Methods

STUDY SITES

Seattle and Vancouver are large port cities in the Pacific Northwest. Although on opposite sides of an international border, they are only 140 miles apart, a three-hour drive by freeway. They share a common geography, climate, and history. Citizens in both cities have attained comparable levels of schooling and have almost identical rates of unemployment. When adjusted to U.S. dollars, the median annual income of a household in Vancouver exceeds that in Seattle by less than $500. Similar percentages of households in both cities have incomes of less than $10,000 (U.S.) annually. Both cities have large white majorities. However, Vancouver has a larger Asian population, whereas Seattle has larger black and Hispanic minorities (Table 1).[10,11] The two communities also share many cultural values and interests. Six of the top nine network television programs in Seattle are among the nine most watched programs in Vancouver.[12,13]

8

Table 1 SOCIOECONOMIC CHARACTERISTICS AND RACIAL AND
ETHNIC COMPOSITION OF THE POPULATIONS IN SEATTLE AND
VANCOUVER

Index	Seattle	Vancouver
1980 Population	493,846	415,220
1985–1986 Population estimate	491,400	430,826
Unemployment rate (%)	5.8	6.0
High-school graduates (%)	79.0	66.0
Median household income		
(U.S. dollars)	16,254	16,681
Households with incomes		
≤$10,000	30.6	28.9
(U.S.) (%)		
Ethnic and racial groups (%)		
White (non-Hispanic)	79.2	75.6
Asian	7.4	22.1
Black	9.5	0.3
Hispanic	2.6	0.5
Native North American	1.3	1.5

FIREARM REGULATIONS

9   Although similar in many ways, Seattle and Vancouver differ
markedly in their approaches to the regulation of firearms (Table
2). In Seattle, handguns may be purchased legally for self-defense
in the street or at home. After a 30-day waiting period, a permit can
be obtained to carry a handgun as a concealed weapon. The recreational use of handguns is minimally restricted.[15]

10   In Vancouver, self-defense is not considered a valid or legal
reason to purchase a handgun. Concealed weapons are not permitted. Recreational uses of handguns (such as target shooting and
collecting) are regulated by the province, and the purchase of a
handgun requires a restricted-weapons permit. A permit to carry a
weapon must also be obtained in order to transport a handgun, and
these weapons can be discharged only at a licensed shooting club.
Handguns can be transported by car, but only if they are stored in
the trunk in a locked box.[16,17]

Table 2 Regulation and Ownership of Firearms and Law-Enforcement Activity in Seattle and Vancouver

	Seattle	Vancouver
Regulations		
Handguns	Concealed-weapons permit is required to carry a gun for self-defense on the street; none is required for self-defense in the home. Registration of handguns is not mandatory for private sales.	Restricted-weapons permit is required for sporting and collecting purposes. Self-defense in the home or street is not legally recognized as a reason for possession of a handgun. Handguns must be registered.
Long guns (rifles, shotguns)	Long guns are not registered.	Firearm-acquisition certificate is required for purchase. Long guns are not registered.
Law enforcement and sentencing		
Additional sentence for commission of a class A felony with a firearm	Minimum of 2 extra years	1 to 14 extra years
Percent of firearm-related homicides that result in police charges (police estimate)	80 to 90%	80 to 90%
Minimum jail sentence for first-degree murder	20 years in prison	25 years in prison (parole is possible after 15 years)
Status of capital punishment	Legal, though no one has been executed since 1963	Abolished
Prevalence of weapons		
Total concealed-weapons permits issued (March 1984 to March 1988)	15,289	—
Total restricted-weapons permits issued (March 1984 to March 1988)	—	4137
Cook's gun prevalence index14	41%	12%

11   Although they differ in their approach to firearm regulations,
both cities aggressively enforce existing gun laws and regulations,
and convictions for gun-related offenses carry similar penalties. For
example, the commission of a class A felony (such as murder or
robbery) with a firearm in Washington State adds a minimum of
two years of confinement to the sentence for the felony.[18] In the
Province of British Columbia, the same offense generally results in
1 to 14 years of imprisonment in addition to the felony sentence.[16]
Similar percentages of homicides in both communities eventually
lead to arrest and police charges. In Washington, under the Sentenc-
ing Reform Act of 1981, murder in the first degree carries a minimum sentence of 20 years of confinement.[19] In British Columbia,
first-degree murder carries a minimum sentence of 25 years, with
a possible judicial parole review after 15 years.[20] Capital punishment
was abolished in Canada during the 1970s.[21] In Washington State,
the death penalty may be invoked in cases of aggravated first-degree
murder, but no one has been executed since 1963.

RATES OF GUN OWNERSHIP

12   Because direct surveys of firearm ownership in Seattle and Vancouver have never been conducted, we assessed the rates of gun ownership indirectly by two independent methods. First, we obtained
from the Firearm Permit Office of the Vancouver police department a count of the restricted-weapons permits issued in Vancouver between March 1984 and March 1988 and compared this
figure with the total number of concealed-weapons permits issued
in Seattle during the same period, obtained from the Office of
Business and Profession Administration, Department of Licensing,
State of Washington. Second, we used Cook's gun prevalence index,
a previously validated measure of intercity differences in the prevalence of gun ownership.[14] This index is based on data from 49 cities
in the United States and correlates each city's rates of suicide and
assaultive homicide involving firearms with survey-based estimates of gun ownership in each city. Both methods indicate that
firearms are far more commonly owned in Seattle than in Vancouver (Table 2).

IDENTIFICATION AND DEFINITION OF CASES

From police records, we identified all the cases of robbery, burglary, 13 and assault (both simple and aggravated) and all the homicides that occurred in Seattle or Vancouver between January 1, 1980, and December 31, 1986. In defining cases, we followed the guidelines of the U.S. Federal Bureau of Investigation's uniform crime reports (UCR).[22] The UCR guidelines define aggravated assault as an unlawful attack by one person on another for the purpose of inflicting severe or aggravated bodily harm. Usually this type of assault involves the actual or threatened use of a deadly weapon. Simple assault is any case of assault that does not involve the threat or use of a deadly weapon or result in serious or aggravated injuries.

A homicide was defined as the willful killing of one human being 14 by another. This category included cases of premeditated murder, intentional killing, and aggravated assault resulting in death. "Justifiable homicide," as defined by the UCR guidelines, was limited to cases of the killing of a felon by a law-enforcement officer in the line of duty or the killing of a felon by a private citizen during the commission of a felony.[22] Homicides that the police, the prosecuting attorney, or both thought were committed in self-defense were also identified and noted separately.

STATISTICAL ANALYSIS

From both Seattle and Vancouver, we obtained annual and cumu- 15 lative data on the rates of aggravated assault, simple assault, robbery, and burglary. Cases of aggravated assault were categorized according to the weapon used. Data on homicides were obtained from the files of the medical examiner or coroner in each community and were supplemented by police case files. Each homicide was further categorized according to the age, sex, and race or ethnic group of the victim, as well as the weapon used.

Population-based rates of simple assault, aggravated assault, rob- 16 bery, burglary, and homicide were then calculated and compared. These rates are expressed as the number per 100,000 persons per year and, when possible, are further adjusted for any differences in the age and sex of the victims. Unadjusted estimates of relative risk and 95 percent confidence intervals were calculated with use of the maximum-likelihood method and are based on Seattle's rate relative to Vancouver's.[23] Age-adjusted relative risks were estimated with use of the Mantel-Haenszel summary odds ratio.[24]

Results

17   During the seven-year study period, the annual rate of robbery in Seattle was found to be only slightly higher than that in Vancouver (relative risk, 1.09; 95 percent confidence interval, 1.08 to 1.12). Burglaries, on the other hand, occurred at nearly identical rates in the two communities (relative risk, 0.99; 95 percent confidence interval, 0.98 to 1.00). During the study period, 18,925 cases of aggravated assault were reported in Seattle, as compared with 12,034 cases in Vancouver. When the annual rates of assault in the two cities were compared for each year of the study, we found that the two communities had similar rates of assault during the first four years of the study. In 1984, however, reported rates of simple and aggravated assault began to climb sharply in Seattle, whereas the rates of simple and aggravated assault remained relatively constant in Vancouver (Fig. 1). This change coincided with the enactment that year of the Domestic Violence Protection Act by the Washington State legislature. Among other provisions, this law required changes in reporting and arrests in cases of domestic violence.[25] It is widely believed that this law and the considerable media attention that followed its passage resulted in dramatic increases in the number of incidents reported and in related enforcement costs in Seattle.[26] Because in Vancouver there was no similar legislative initiative requiring police to change their reporting methods, we restricted our comparison of the data on assaults to the first four years of our study (1980 through 1983) (Fig. 1).

18   During this four-year period, the risk of being a victim of simple assault in Seattle was found to be only slightly higher than that in Vancouver (relative risk, 1.18; 95 percent confidence interval, 1.15 to 1.20). The risk of aggravated assault in Seattle was also only slightly higher than in Vancouver (relative risk, 1.16; 95 percent

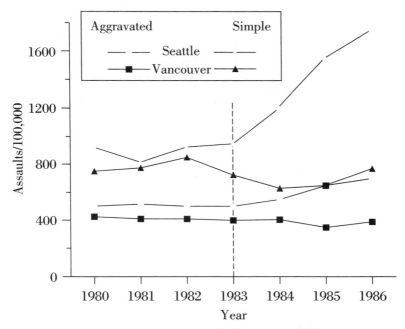

Figure 1 Rates of aggravated and simple assault in Seattle and Vancouver, 1980 through 1986. The dotted line indicates the passage of the Domestic Violence Protection Act in Washington State in 1984.

confidence interval, 1.12 to 1.19). However, when aggravated assaults were subdivided by the type of weapon used and the mechanism of assault, a striking pattern emerged. Although both cities reported almost identical rates of aggravated assault involving knives, other dangerous weapons, or hands, fists, and feet, firearms were far more likely to have been used in cases of assault in Seattle than in Vancouver (Table 3). In fact, all the difference in the relative risk of aggravated assault between these two communities was due to Seattle's 7.7-fold higher rate of assaults involving firearms (Fig. 2).

Over the whole seven-year study period, 388 homicides occurred in Seattle (11.3 per 100,000 person-years). In Vancouver, 204 homicides occurred during the same period (6.9 per 100,000 person-years). After adjustment for differences in age and sex between the

19

Table 3 ANNUAL CRUDE RATES AND RELATIVE RISKS OF AGGRAVATED ASSAULT, SIMPLE ASSAULT, ROBBERY, BURGLARY, AND HOMICIDE IN SEATTLE AND VANCOUVER, 1980 THROUGH 1986*

Crime	Period	Seattle	Vancouver	Relative Risk	95% CI
		(no./100,000)			
Robbery	1980–1986	492.2	450.9	1.09	1.08–1.12
Burglary	1980–1986	2952.7	2985.7	0.99	0.98–1.00
Simple assault	1980–1983	902	767.7	1.18	1.15–1.20
Aggravated assault	1980–1983	486.5	420.5	1.16	1.12–1.19
Firearms		87.9	11.4	7.70	6.70–8.70
Knives		78.1	78.9	0.99	0.92–1.07
Other		320.6	330.2	0.97	0.94–1.01
Homicides	1980–1986	11.3	6.9	1.63	1.38–1.93
Firearms		4.8	1.0	5.08	3.54–7.27
Knives		3.1	3.5	0.90	0.69–1.18
Other		3.4	2.5	1.33	0.99–1.78

*CI denotes confidence interval. The "crude rate" for these crimes is the number of events occurring in a given population over a given time period. The relative risks shown are for Seattle in relation to Vancouver

populations, the relative risk of being a victim of homicide in Seattle, as compared with Vancouver, was found to be 1.63 (95 percent confidence interval, 1.28 to 2.08). This difference is highly unlikely to have occurred by chance.

20   When homicides were subdivided by the mechanism of death, the rate of homicide by knives and other weapons (excluding firearms) in Seattle was found to be almost identical to that in Vancouver (relative risk, 1.08; 95 percent confidence interval, 0.89 to 1.32) (Fig. 3). Virtually all of the increased risk of death from homicide in Seattle was due to a more than fivefold higher rate of homicide by firearms (Table 3). Handguns, which accounted for roughly 85 percent of the homicides involving firearms in both communities, were 4.8 times more likely to be used in homicides in Seattle than in Vancouver.

21   To test the hypothesis that the higher rates of homicide in Seattle might be due to more frequent use of firearms for self-protection, we examined all the homicides in both cities that were ruled

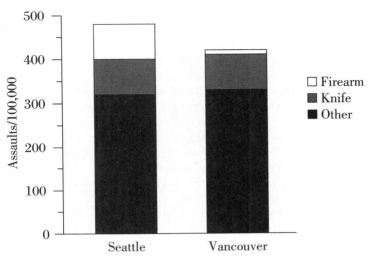

Figure 2   Annual rates of aggravated assault in Seattle and Vancouver, 1980 through 1983, according to the Weapon Used. "Other" includes blunt instruments, other dangerous weapons, and hands, fists, and feet.

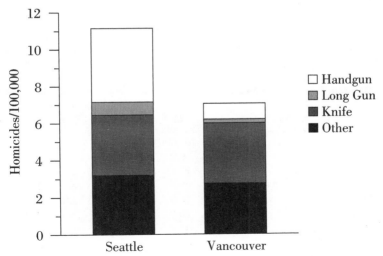

Figure 3   Annual rates of homicide in Seattle and Vancouver, 1980 through 1986, according to the Weapon Used. "Other" includes blunt instruments, other dangerous weapons, and hands, fists, and feet.

"legally justifiable" or were determined to have been committed in self-defense. Thirty-two such homicides occurred during the study period, 11 of which involved police intervention. After the exclusion of justifiable homicide by police, 21 cases of homicide by civilians acting in self-defense or in other legally justifiable ways remained, 17 of which occurred in Seattle and 4 of which occurred in Vancouver (relative risk, 3.64; 95 percent confidence interval, 1.32 to 10.06). Thirteen of these cases (all of which occurred in Seattle) involved firearms. The exclusion of all 21 cases (which accounted for less than 4 percent of the homicides during the study interval) had little overall effect on the relative risk of homicide in the two communities (age- and sex-adjusted relative risk, 1.57; 95 percent confidence interval, 1.22 to 2.01).

22   When homicides were stratified by the race or ethnic group of the victim, a complex picture emerged (Table 4). The homicide rates in Table 4 were adjusted for age to match the 1980 U.S. population. This technique permits fairer comparisons among racial and ethnic groups with differing age compositions in each city. The relative risk for each racial or ethnic group, however, was estimated with use of the Mantel-Haenszel summary odds ratio.[24] This

Table 4 ANNUAL AGE-ADJUSTED HOMICIDE RATES AND RELATIVE RISKS OF DEATH BY HOMICIDE IN SEATTLE AND VANCOUVER, 1980 THROUGH 1986, ACCORDING TO THE RACE OR ETHNIC GROUP OF THE VICTIM*

Race or Ethnic Group	Seattle	Vancouver	Relative Risk	95% CI
	(no./100,000)			
White (non-Hispanic)	6.2	6.4	1	0.8–1.2
Asian	15.0	4.1	3.5	2.1–5.7
Excluding Wah Mee murders	9.5	—	2.3	1.4–4.0
Black	36.6	9.5	2.8	0.4–20.4
Hispanic	26.9	7.9	5	0.7–34.3
Native American	64.9	71.3	0.9	0.5–1.5

*CI denotes confidence interval. The relative risks shown are for Seattle in relation to Vancouver.

method, in effect, uses a different set off weights for the various age strata, depending on the distribution of persons among the age strata for that racial or ethnic group only. Hence, these estimates of relative risk differ slightly from a simple quotient of the age-adjusted rates.

Whereas similar rates of death by homicide were noted for whites in both cities, Asians in Seattle had higher rates of death by homicide than their counterparts in Vancouver. This difference persisted even after the exclusion of the 13 persons who died in the Wah Mee gambling club massacre in Seattle in 1983. Blacks and Hispanics in Seattle had higher relative risks of death by homicide than blacks and Hispanics in Vancouver, but the confidence intervals were very wide, given the relatively small size of both minorities in Vancouver. Only one black and one Hispanic were killed in Vancouver during the study period. Native Americans had the highest rates of death by homicide in both cities.

Discussion

Previous studies of the effectiveness of gun control have generally compared rates of homicide in nations with different approaches to the regulation of firearms.[7] Unfortunately, the validity of these studies has been compromised by the large number of confounding factors that characterize national groups. We sought to circumvent this limitation by focusing our analysis on two demographically comparable and physically proximate cities with markedly different approaches to handgun control. In many ways, these two cities have more in common with each other than they do with other major cities in their respective countries. For example, Seattle's homicide rate is consistently half to two thirds that reported in cities such as Chicago, Los Angeles, New York, and Houston,[4] whereas Vancouver experiences annual rates of homicide two to three times higher than those reported in Ottawa, Toronto, and Calgary (Canadian Centre for Justice Statistics, Homicide Program, Ottawa: unpublished data).

In order to exclude the possibility that Seattle's higher homicide rate may be explained by higher levels of criminal activity or

23

24

25

aggressiveness in its population, we compared the rates of burglary, robbery, simple assault, and aggravated assault in the two communities. Although we observed a slightly higher rate of simple and aggravated assault in Seattle, these differences were relatively small—the rates in Seattle were 16 to 18 percent higher than those reported in Vancouver during a period of comparable case reporting. Virtually all of the excess risk of aggravated assault in Seattle was explained by a sevenfold higher rate of assaults involving firearms. Despite similar rates of robbery and burglary and only small differences in the rates of simple and aggravated assault, we found that Seattle had substantially higher rates of homicide than Vancouver. Most of the excess mortality was due to an almost fivefold higher rate of murders with handguns in Seattle.

26   Critics of handgun control have long claimed that limiting access to guns will have little effect on the rates of homicide, because persons who are intent on killing others will only work harder to acquire a gun or will kill by other means.[7,27] If the rate of homicide in a community were influenced more by the strength of intent than by the availability of weapons, we might have expected the rate of homicides with weapons other than guns to have been higher in Vancouver than in Seattle, in direct proportion to any decrease in Vancouver's rate of firearm homicides. This was not the case. During the study interval, Vancouver's rate of homicides with weapons other than guns was not significantly higher than that in Seattle, suggesting that few would-be assailants switched to homicide by other methods.

27   Ready access to handguns has been advocated by some as an important way to provide law-abiding citizens with an effective means to defend themselves.[27–29] Were this true, we might have expected that much of Seattle's excess rate of homicides, as compared with Vancouver's, would have been explained by a higher rate of justifiable homicides and killings in self-defense by civilians. Although such homicides did occur at a significantly higher rate in Seattle than in Vancouver, these cases accounted for less than 4 percent of the homicides in both cities during the study period. When we excluded cases of justifiable homicide or killings in self-defense by civilians from our calculation of relative risk, our results were almost the same.

28   It also appears unlikely that differences in law-enforcement

activity accounted for the lower homicide rate in Vancouver. Suspected offenders are arrested and cases are cleared at similar rates in both cities. After arrest and conviction, similar crimes carry similar penalties in the courts in Seattle and Vancouver.

We found substantial differences in the risk of death by homicide according to race and ethnic group in both cities. In the United States, blacks and Hispanics are murdered at substantially higher rates than whites.[2] Although the great majority of homicides in the United States involve assailants of the same race or ethnic group, current evidence suggests that socioeconomic status plays a much greater role in explaining racial and ethnic differences in the rate of homicide than any intrinsic tendency toward violence.[2,30,31] For example, Centerwall has shown that when household crowding is taken into account, the rate of domestic homicide among blacks in Atlanta, Georgia, is no higher than that of whites living in similar conditions.[32] Likewise, a recent study of childhood homicide in Ohio found that once cases were stratified by socioeconomic status, there was little difference in race-specific rates of homicide involving children 5 to 14 years of age.[33]

Since low-income populations have higher rates of homicide, socioeconomic status is probably an important confounding factor in our comparison of the rates of homicide for racial and ethnic groups. Although the median income and the overall distribution of household incomes in Seattle and Vancouver are similar, the distribution of household incomes by racial and ethnic group may not be the same in Vancouver as in Seattle. For example, blacks in Vancouver had a slightly higher mean income in 1981 than the rest of Vancouver's population (Statistics Canada, 1981 Census Custom Tabulation: unpublished data). In contrast, blacks in Seattle have a substantially lower median income than the rest of Seattle's population.[34] Thus, much of the excess risk of homicide among blacks in Seattle, as compared with blacks in Vancouver, may be explained by their lower socioeconomic status. If, on the other hand, more whites in Vancouver have low incomes than whites in Seattle, the higher risk of homicide expected in this low-income subset may push the rate of homicide among whites in Vancouver higher than that for whites in Seattle. Unfortunately, neither hypothesis can be tested in a quantitative fashion, since detailed information about household incomes according to race is not available for Vancouver.

31   Three limitations of our study warrant comment. First, our measures of the prevalence of firearm ownership may not precisely reflect the availability of guns in the two communities. Although the two measures we used were derived independently and are consistent with the expected effects of gun control, their validity as indicators of community rates of gun ownership has not been conclusively established. Cook's gun prevalence index has been shown to correlate with data derived from national surveys, but it has not been tested for accuracy in cities outside the United States. Comparisons of concealed-weapons permits in Seattle with restricted-weapons permits in Vancouver are probably of limited validity, since these counts do not include handguns obtained illegally. In fact, the comparison of permit data of this sort probably substantially underestimates the differences between the communities in the rate of handgun ownership, since only a fraction of the handguns in Seattle are purchased for use as concealed weapons, whereas all legal handgun purchases in Vancouver require a restricted-weapons permit. Still, these indirect estimates of gun ownership are consistent with one another, and both agree with prior reports

References

1. Homicide surveillance: 1970–78. Atlanta: Centers for Disease Control, September, 1983.
2. Homicide surveillance: high risk racial and ethnic groups—blacks and Hispanics, 1970 to 1983. Atlanta: Centers for Disease Control, November, 1986.
3. Baker SP, O'Neill B, Karpf RS. The injury fact book. Lexington, Mass.: Lexington Books, 1984.
4. Department of Justice, Federal Bureau of Investigation. Crime in the United States (Uniform Crime Reports). Washington, D.C.: Government Printing Office, 1986.
5. Baker SP. Without guns, do people kill people? Am J Public Health 1985; 75:587–8.
6. Hedeboe J, Charles AV, Nielsen J, et al. Interpersonal violence: patterns in a Danish community. Am J Public Health 1985; 75:651–3.
7. Wright J, Rossi P, Daly K, Weber-Burdin E. Weapons, crime and violence in America: a literature review and research agenda. Washington, D.C.: Department of Justice, National Institute of Justice, 1981.

8. Weiss JMA. Gun control: a question of public/mental health? J Oper Psychiatr 1981; 12:86–8.

9. Bruce-Briggs B. The great American gun war. Public Interest 1976; 45:37–62.

10. Bureau of Census. 1980 Census of population, Washington. Washington, D.C.: Government Printing Office, 1981.

11. Statistics Canada: 1981 census of Canada, Vancouver, British Columbia. Ottawa, Ont.: Minister of Supply and Services, 1983.

12. Seattle local market T.V. ratings, 1985–86. (Based on Arbitron television ratings.) Provided by KING TV, Seattle, Washington.

13. Vancouver local market T.V. ratings, 1985–86. Provided by Bureau of Broadcast Measurement, Toronto.

14. Cook PJ. The role of firearms in violent crime. In: Wolfgang M, ed. Criminal violence. Beverly Hills, Calif.: Sage, 1982:236–90.

15. Revised Code of State of Washington. RCW chapter 9.41.090, 9.41.095, 9.41.070, 1986.

16. Criminal Code of Canada. Firearms and other offensive weapons. Martin's Criminal Code of Canada, 1982. Part II.1 (Sections 81-016.9, 1982).

17. *Idem.* Restricted Weapons and Firearm Control Regulations Sec. 106.2 (11); Amendment Act, July 18, 1977, 1982.

18. Revised Code of State of Washington, Sentence Reform Act Chapter 9 94A.125.1980.

19. Revised Code of State of Washington. Murder I, 9A.32.040.1984.

20. Criminal Code of Canada. Application for judicial review sentence of life imprisonment, 1988 Part XX 669–67, 1(1).

21. *Idem.* Act to Amend Criminal Code B.11 C84, 1976.

22. Department of Justice, Federal Bureau of Investigation. Uniform crime reporting handbook. Washington, D.C.: Government Printing Office, 1984.

23. Rothman KJ, Boice JD Jr. Epidemiologic analysis with a programmable calculator. Boston: Epidemiology Resources, 1982.

24. Armitage P, Berry G. Statistical methods in medical research. 2nd ed. Oxford: Blackwell, 1987.

25. Revised Code of State of Washington. RCW Chapter 10.99.010-.100, 1984.

26. Seattle Police Department. Inspectional service division report, domestic violence arrest costs: 1984–87, Seattle, 1986.

27. Drooz RB. Handguns and hokum: a methodological problem. JAMA 1977; 238:43–5.

28. Copeland AR. The right to keep and bear arms—a study of civilian homicides committed against those involved in criminal acts in metropolitan Dade County from 1957 to 1982. J Forensic Sci 1984; 29:584–90.

29. Kleck G. Crime control through the private use of armed force. Soc Probl 1988; 35:1–21.

30. Loftin C, Hill RH. Regional subculture and homicide: an examination of the Gastil-Hackney thesis. Am Sociol Rev 1974; 39:714–24.

31. Williams KR. Economic sources of homicide: reestimating the effects of poverty and inequality. Am Sociol Rev 1984; 49:283–9.

32. Centerwall BS. Race, socioeconomic status, and domestic homicide, Atlanta, 1971–72. Am J Public Health 1984; 74:813–5.
33. Muscat JE. Characteristics of childhood homicide in Ohio, 1974–84. Am J Public Health 1988; 78:822–4.
34. Seattle City Government. General social and economic characteristics, city of Seattle: 1970–1980. Planning research bulletin no. 45. Seattle: Department of Community Development, 1983.
35. Newton G, Zimring F. Firearms and violence in American life: a staff report to the National Commission on the Causes and Prevention of Violence. Washington, D.C.: Government Printing Office, 1969.

STUDY QUESTIONS

1. Sloan, et al., conclude that ease of access to handguns is the major causal factor relevant to explaining homicide rates. Which of Mill's Methods does their overall method of argument employ?

2. In paragraph 29, Sloan, et al., hypothesize that socioeconomic factors (and not intrinsic racial differences) are also causally relevant to explaining homicide rates. This hypothesis is based on the results summarized in their Table 4. Which of Mill's Methods leads them to this hypothesis?

3. Notice that while homicide rates are much higher in Seattle than in Vancouver, the unemployment and income statistics given in Table 1 show Seattle and Vancouver to be very similar economically. Why, in paragraph 30, do Sloan, et al., think that these statistics should not lead us to reject the hypothesis that economic factors are relevant?

4. According to Table 1, residents of Vancouver are socioeconomically similar to residents of Seattle. Also according to Table 1, whites make up about the same percentage of the population in Vancouver as they do in Seattle. And according to Table 4, homicide rates for whites in Vancouver and Seattle are nearly identical. Yet whites in Seattle have much easier access to firearms than do whites in Vancouver. Using the negative Method of Difference, what conclusion follows? How do you think Sloan, et al., would respond to this?

5. In his essay against banning handguns (p. 281), did Kates raise any arguments that Sloan, et al., have not dealt with in this essay?

Reference Chapters: 15, 17

DAVID RUBINSTEIN

Don't Blame Crime on Joblessness

David Rubinstein is a professor of sociology at the University of Illinois at Chicago. The following selection on the relationship between crime and unemployment first appeared in the op-ed page of The Wall Street Journal.

The California Assembly Special Committee on the Los Angeles   1
crisis recently released its findings on the riots of last spring. Unsurprisingly, the report echoes the Kerner Commission of a generation ago by emphasizing lack of economic opportunity as a major cause of the riots and the high crime rates in South Central Los Angeles.

The coincidence of crime and unemployment in places like South   2
Central Los Angeles seems to confirm their connection. And it makes sense motivationally. Surely an absence of employment can make crime an attractive option, and so enhanced job opportunities

[David Rubinstein, "Don't Blame Crime on Joblessness." *The Wall Street Journal*, November 9, 1992, op-ed page.]

ought to make it less so. University of Chicago economist Gary S. Becker just won the Nobel Prize for this sort of reasoning.

3   But there are profound anomalies in this analysis. First, the place of crime in the life cycle is odd. One would think that limited job options would mean more to a man approaching 30 than to a teen-ager. But conviction rates for men between 25 and 30 are about one-third the rates for boys between 14 and 16. Similarly, a man with a family faces more urgent economic imperatives than a single man, and yet his inclination to crime is far less. It is noteworthy that women, despite various economic barriers, are invariably less prone to crime than men.

4   Also, it is hard to see crimes like rape, drug addiction, most homicides and assaults as substitutes for legitimate employment. Even profit-oriented crime is often of doubtful economic benefit. The take in most petty street crimes is so low that, even with a small chance of arrest for any single crime, a perpetrator will likely be jailed before he equals a year's income from a minimum wage job.

5   With a little ingenuity, the economic interpretation can be stretched to "explain" crimes that lack economic sense. While stabbing someone in a bar fight, using drugs or setting fire to a store are hardly substitutes for gainful employment, such crimes might be interpreted as "ultimately" reflecting the frustrations of blocked opportunities.

6   But all such theories founder on a striking fact: the nearly invisible relationship between unemployment and crime rates. Charting homicide since 1900 reveals two peaks. The first is in 1933. This represents the crest of a wave that began in 1905, continued through the prosperous '20s and then began to *decline* in 1934 as the Great Depression was deepening. Between 1933 and 1940, the murder rate dropped nearly 40%. Property crimes reveal a similar pattern.

7   Between 1940 and 1960 the homicide rate remained relatively stable. In the early '60s, a sharp increase began that peaked in 1974, when the murder rate was more than double that of the late 1950s, and far higher than it had been in the depths of the Depression. Between 1963 and 1973 homicides in New York City tripled. Again, property and most other forms of crime followed a similar pattern.

8   The cause of this remarkable increase in crime certainly was not unemployment—which was, by contemporary standards, enviably

low. In 1961, the unemployment rate was 6.6% and the crime rate was 1.9 per 1,000. By 1969, unemployment had dropped to 3.4% while the crime rate nearly doubled to 3.7 per 1,000. The incidence of robbery nearly tripled. Interestingly, the recession of 1980 to 1982 was accompanied by a small but clearly discernible drop in crime. As the economy revived, so did the crime rate.

These patterns are well known to criminologists. A review of   9 several studies by Thomas Orsagh concluded that "unemployment may affect the crime rate, but even if it does, its general effect is too slight to be measured." Another survey by Richard Freeman concluded that the relationship is so weak that, if unemployment were cut by 50%, the crime rate would drop by only 5%. Some criminologists seriously entertain the thesis that crime, like any other form of "business" activity, turns up in good times.

Despite this evidence, the idea that crime can be substantially cut   10 by enhanced employment opportunities remains deeply entrenched, even in the social sciences. Ours is a materialistic culture. We believe that people are driven by calculations of economic gain and that money can solve a host of social problems. But human motivation is far more diverse, and often darker, than this. Just as money spent on health care can do little to counter the effects of destructive life styles, and money spent on schools cannot overcome a lack of motivation to learn, pouring money into America's inner cities to enhance employment opportunities will do little to make them safer.

When considering what to do about crime in places like South   11 Central Los Angeles, it is worth recalling the relationship between crime and economic deprivation in a different part of California at a different time: "During the 1960s, one neighborhood in San Francisco had the lowest income, the highest unemployment rate, the highest proportion of families with incomes under $4,000 a year, the least educational attainment, the highest tuberculosis rate, and the highest proportion of substandard housing. . . . That neighborhood was called Chinatown. Yet in 1965, there were only five persons of Chinese ancestry committed to prison in the entire state of California."

This quote, taken from *Crime and Human Nature* by James Q.   12 Wilson and Richard Herrnstein, suggests that economic theories tell us more about our misunderstandings of human motivation than about the causes of crime. It also suggests that policy planners would rather speak of factors that are within the reach of government programs than those, like weak families and a culture that fails to restrain, that are truly related to crime.

STUDY QUESTIONS

1. In the first two paragraphs, Rubinstein sketches the argument that he disagrees with. Rewrite that argument into standard form.
2. Rubinstein presents three deductive arguments in paragraph 3—one based on comparing men between ages 25 and 30 with boys between ages 14 and 16; one based on comparing men with families with single men; and one based on comparing women with men. All of these arguments contain assumed premises. Supply the assumed premises and reconstruct the arguments into standard form.
3. In paragraphs 6 to 8, Rubinstein presents a series of statistics correlating the homicide rate and unemployment rates. In concluding that unemployment does not lead to crime, which of Mill's Methods is he using?
4. At the end of his article, Rubinstein hypothesizes that "weak families and a culture that fails to restrain" are actual causes of crime. In support of this hypothesis he offers two analogies in paragraph 10, and in paragraph 11 he mentions San Francisco's Chinatown in the 1960s. How much support do these give to his hypothesis?

Reference Chapters: 7, 10, 15, 18

MICHAEL HARRINGTON

Defining Poverty

Michael Harrington, born in 1928 in St. Louis, was a leading American socialist. He was a professor of political science at Queens College, City University of New York, and the author of The Other America, *from which the following excerpt is taken.*

In the nineteenth century, conservatives in England used to argue   1
against reform on the grounds that the British worker of the time
had a longer life expectancy than a medieval nobleman.

This is to say that a definition of poverty is, to a considerable   2
extent, a historically conditioned matter. Indeed, if one wanted to
play with figures, it would be possible to prove that there are no
poor people in the United States, or at least only a few whose plight
is as desperate as that of masses in Hong Kong. There is starvation
in American society, but it is not a pervasive social problem as it is
in some of the newly independent nations. There are still Americans
who literally die in the streets, but their numbers are comparatively
small.

This abstract approach toward poverty in which one compares   3
different centuries or societies has very real consequences. For the
nineteenth century British conservative, it was a way of ignoring
the plight of workers who were living under the most inhuman
conditions. The twentieth century conservative would be shocked

[Michael Harrington, excerpt from the Appendix to *The Other America: Poverty in the United States.* New York: Pelican Books, 1971, pp. 187–190.]

and appalled in an advanced society if there were widespread conditions like those of the English cities a hundred years ago. Our standards of decency, of what a truly human life requires, change, and they should.

4   There are two main aspects of this change. First, there are new definitions of what man can achieve, of what a human standard of life should be. In recent times this has been particularly true since technology has consistently broadened man's potential: it has made a longer, healthier, better life possible. Thus, in terms of what is technically possible, we have higher aspirations. Those who suffer levels of life well below those that are possible, even though they live better than medieval knights or Asian peasants, are poor.

5   Related to this technological advance is the social definition of poverty. The American poor are not poor in Hong Kong or in the sixteenth century; they are poor here and now, in the United States. They are dispossessed in terms of what the rest of the nation enjoys, in terms of what the society could provide if it had the will. They live on the fringe, the margin. They watch the movies and read the magazines of affluent America, and these tell them that they are internal exiles.

6   To some, this description of the feelings of the poor might seem to be out of place in discussing a definition of poverty. Yet if this book indicates anything about the other America, it is that this sense of exclusion is the source of a pessimism, a defeatism that intensifies the exclusion. To have one bowl of rice in a society where all other people have half a bowl may well be a sign of achievement and intelligence; it may spur a person to act and to fulfill his human potential. To have five bowls of rice in a society where the majority have a decent, balanced diet is a tragedy.

7   This point can be put another way in defining poverty. One of the consequences of our new technology is that we have created new needs. There are more people who live longer. Therefore they need more. In short, if there is technological advance without social advance, there is, almost automatically, an increase in human misery, in impoverishment.

8   And finally, in defining poverty one must also compute the social cost of progress. One of the reasons that the income figures show fewer people today with low incomes than twenty years ago is that more wives are working now, and family income has risen as a

result. In 1940, 15 per cent of wives were in the labor force; in 1957 the figure was 30 per cent. This means that there was more money and, presumably, less poverty.

Yet a tremendous growth in the number of working wives is an 9 expensive way to increase income. It will be paid for in terms of the impoverishment of home life, of children who receive less care, love, and supervision. This one fact, for instance, might well play a significant role in the problems of the young in America. It could mean that the next generation, or a part of it, will have to pay the bill for the extra money that was gained. It could mean that we have made an improvement in income statistics at the cost of hurting thousands and hundreds of thousands of children. If a person has more money but achieves this through mortgaging the future, who is to say that he or she is no longer poor?

It is difficult to take all these imponderables together and to 10 fashion them into a simple definition of poverty in the United States. Yet this analysis should make clear some of the assumptions that underlie the assertions in this book:

Poverty should be defined in terms of those who are denied the 11 minimal levels of health, housing, food, and education that our present stage of scientific knowledge specifies as necessary for life as it is now lived in the United States.

Poverty should be defined psychologically in terms of those 12 whose place in the society is such that they are internal exiles who, almost inevitably, develop attitudes of defeat and pessimism and who are therefore excluded from taking advantage of new opportunities.

Poverty should be defined absolutely, in terms of what man and 13 society could be. As long as America is less than its potential, the nation as a whole is impoverished by that fact. As long as there is the other America, we are, all of us, poorer because of it.

STUDY QUESTIONS

1. In paragraphs 11–13, Harrington offers three guidelines for defining poverty. Summarize this information into a single concise definition.

2. Central to Harrington's psychological dimension of poverty is his notion of an "internal exile." Does he use this concept literally or metaphorically? If he uses it metaphorically, does it violate the rule that prohibits metaphorical definitions?

3. In paragraph 9, Harrington suggests that an increase in the number of working wives "will be paid for in terms of the impoverishment of home life." What evidence does he provide for this?

4. How would Harrington respond to the following objection: "Poverty should not be defined relatively, because poverty is a matter of basic needs not being met, and while people's *expectations* have increased, their basic *needs* have stayed the same"?

5. Does Harrington's argument depend on an assumed premise to the effect that one's material circumstances determine one's psychological outlook? Suppose, for instance, that two individuals have the same minimal amount of money and education, but while one feels optimistic about his or her future, the other feels pessimistic. Would Harrington say they are equally poor? Do you agree?

Reference Chapters: 3, 5

ACKNOWLEGMENTS

Caroline Bird: "College Is a Waste of Time and Money." From *The Case Against College* by Caroline Bird. Copyright © 1975 by Caroline Bird. Reprinted by permission of the author.

T. G. R. Bower: "The Visual World of Infants." From *Scientific American,* December, 1966. Reprinted with permission. Copyright © 1966 by Scientific American, Incorporated. All rights reserved.

Baruch Brody: "Fetal Humanity and Brain Function." From "Fetal Humanity and Theory of Essentialism." From Robert Baker and Frederick Elliston, eds., *Philosophy and Sex* (Buffalo, N.Y.: Prometheus Books). Copyright © 1975 by Prometheus Books. Reprinted by permission of the publisher.

Kenneth Burke: "Classifying Proverbs." From *The Philosophy of Literary Form: Studies in Symbolic Action,* Second Edition, by Kenneth Burke. Copyright © 1967 by Kenneth Burke. Reprinted by permission of The Louisiana State University Press.

E. K. Daniel: "Two Proofs of God's Existence." From "A Defense of Theism." *Philosophy: The Basic Issues,* Third Edition, by E. D. Klemke, A. David Kline, and Robert Hollinger, eds. Copyright © 1990. Reprinted by permission of St. Martin's Press, Incorporated.

Alan Dershowitz: "Shouting Fire!" From the *Atlantic Monthly,* January, 1989. Reprinted by permission of the author.

Arthur Conan Doyle: "The Adventure of Black Peter." From *The Complete Sherlock Holmes,* Vol. II (New York: Doubleday & Co.). Reprinted by permission of Andrea Plunkett.

John Enright: "What Is Poetry?" From *Objectively Speaking,* Autumn, 1989. Reprinted by permission of the author.

Peter Farb: "Children's Insults." From *Word Play: What Happens When People Talk* by Peter Farb. Copyright © 1973 by Peter Farb. Reprinted by permission of Alfred A. Knopf, Incorporated.

Sigmund Freud: "Human Nature Is Inherently Bad." From *Civilization and Its Discontents* by Sigmund Freud, translated by James Strachey. Copyright © 1961 by James Strachey. Copyright renewed 1989. Reprinted by permission of W. W. Norton and Company, Inc.

Gorgias: "Encomium of Helen," translated by Douglas Maurice MacDowell. Copyright © 1982. Reprinted with the permission of Routledge/Bristol Classical Press.

Stephen Jay Gould: "Sex, Drugs, Disasters, and the Extinction of Dinosaurs." From *The Flamingo's Smile: Reflections in Natural History* by Stephen Jay Gould. Copyright © 1985 by Stephen Jay Gould. Reprinted by permission of W. W. Norton and Company, Inc.

Michael Harrington: "Defining Poverty." From *The Other America: Poverty in the United States* by Michael Harrington. Copyright © 1962, 1969 by Michael Harrington. Reprinted with the permission of Macmillan Publishing Company.

Henry Hazlitt: "Who's 'Protected' by Tariffs?" From Chapter XI of *Economics in One Lesson* by Henry Hazlitt. Copyright © 1962, 1979 by Henry Hazlitt. Reprinted by permission of Crown Publishers, Inc.

E. D. Hirsch, Jr.: "Literacy and Cultural Literacy." From *Cultural Literacy* by E. D. Hirsch, Jr. Copyright © 1987 by Houghton Mifflin Company. Reprinted by permission of Houghton Mifflin Company. All rights reserved.

Don B. Kates, Jr.: "Handgun Bans—Facts to Fight With." From *Guns and Ammo Annual,* 1984, pp. 4, 6, 8–11. Los Angeles: Peterson, 1984. Reprinted by permission of *Guns and Ammo.*

Michael Kinsley: "Bias and Baloney." From *The New Republic.* Reprinted by permission of *The New Republic,* copyright © 1992, The New Republic, Incorporated.

Michael Levin: "The Case for Torture." From *Newsweek,* June 7, 1982. Reprinted by permission of the author.

Lysias: "On the Murder of Eratosthenes: Defence." From *On the Murder of Eratosthenes,* translated by W. R. M. Lamb. Copyright © 1957 by Harvard University Press. Reprinted with permission of the publishers and the Loeb Classical Library from Lysias: *On the Murder of Eratosthenes,* translated by W. R. M. Lamb, Cambridge, Mass.: Harvard University Press, 1957.

Niccolò Machiavelli: "On Cruelty and Clemency: Whether It Is Better to Be Loved or Feared." From *The Prince* by Niccolò Machiavelli, A Norton Critical Edition,

Shelia Tobias: "Mathematics and Sex." From *Overcoming Math Anxiety,* by Shelia Tobias. Copyright © 1978 by Shelia Tobias. Reprinted by permission of W. W. Norton and Company, Incorporated.

Laurence H. Tribe: "Opposition to Abortion Is Not Based on Alleged Rights to Life." From *Abortion: The Clash of Absolutes* by Laurence H. Tribe. Copyright © 1990 by Laurence H. Tribe. Reprinted by permission of W. W. Norton and Company, Inc.

Mary Anne Warren: "On the Moral and Legal Status of Abortion, Part II." From *The Monist,* January, 1973. Copyright © 1973, *The Monist,* La Salle, Illinois 61301. Reprinted by permission.